Wellbeing and Self-Tran:
Landsc

Rebecca Crowther

Wellbeing and Self-Transformation in Natural Landscapes

Rebecca Crowther
University of Edinburgh
Edinburgh, UK

ISBN 978-3-030-07387-9 ISBN 978-3-319-97673-0 (eBook)
https://doi.org/10.1007/978-3-319-97673-0

Softcover re-print of the Hardcover 1st edition 2019

Cover illustration: Fatima Jamadar

This Palgrave Macmillan imprint is published by the registered company Springer Nature Switzerland AG
The registered company address is: Gewerbestrasse 11, 6330 Cham, Switzerland

For Rodica

Acknowledgements

I want to thank all my individual informants, the organisations, and the groups of individuals who allowed me into their spaces. You give this book life and myself so many memories.

Contents

List of Figures

1

Introduction: The Phenomenon

As children, my sister and I often played outside, as many children in Scotland may have in the 1980s. We built gang huts out of branches, stones, and debris. Through our stories, shared experience, and collaboration in making our secret hideaways, we made the spaces in which we interacted with friends, our place. Through touching, smelling, climbing, poking around, eating, and breaking bones (!) outside, we developed a part of who we were to become. We would gather tadpoles from beneath an old sewage tank, where the concrete basin had filled with rain water. The old tank sat amongst the trees in a woodland surrounding the old Victoria Hospital in Edinburgh, behind our home—most likely now a space out of bounds to children. We would watch the frogs lazily stretching down and beneath the murky water, their springy leg the last thing to plop in. We would rush tadpoles home, to our mother's despair, to keep in the bath. We had jars with caterpillars in, making sure they had foliage to chomp on. We would catch bees by flowering bushes in jars too, listen to them buzz frantically before letting them free as we squealed.

We would hunt for things, beasties, and passage ways, and we'd leap out of trees. It would turn from daylight to dusk and only then would we come inside. We always found somewhere to play. But at some stage, I can't specify when, we stopped looking.

R. Crowther, *Wellbeing and Self-Transformation in Natural Landscapes*,
https://doi.org/10.1007/978-3-319-97673-0_1

When I was 18 years old, I moved to London to study and was rarely outside of the city centre. The most that I engaged with anything remotely 'natural' was in growing herbs in my window or perhaps sitting in a large, busy public park on a summer's eve with friends and a glass of wine. Years in London, as a student and waitress, went by with very little thought. It was all late nights out and cramming to finish papers. I was simply getting on with 'stuff,' as people do, without tapping into how I was, mentally or physically. At 22, no longer a student and needing to find work, I moved to Birmingham City Centre, where the opportunities or chance engagements with any type of green space were minimal. I was working 18-hour shifts in an events venue, living next to one of the UK's biggest shopping centres and living in a loop: work, pub, home, work, pub, home. I found myself exhausted, run down, and suffering badly with an anxiety disorder that had plagued me since I was a teenager. A good friend at the time insisted that we get out of the city. On the outskirts of Birmingham there was a vast park, Sutton Coldfield, with woodland and lake, and we were going to spend the day there. We wandered, paddled, sat, and talked. Within one month I had handed in my notice to leave at work and was heading back to Scotland.

Looking back, I feel that through our engagement with this childhood exploration and play, my sister and I, alongside our gang of neighbourhood friends, found a temporary sense of belonging. It is due to these encounters that I associate the outdoors with my own personal narrative. These days brought with them a sense of collective discovery, which I continue to see as vital as an adult. It is experiences such as these that have formed my interest in such a topic. As a person who has engaged with these outdoor landscapes due to a desire linked to a childhood spent in urban green space and visiting family in remote rural areas, there is a fond interest in spending time within these spaces. As an adult, I *choose* to engage with these spaces. I therefore position myself, as a researcher, as someone who deliberately engages with natural environments due to my own perceived benefits, regardless of their intangibility. Living within urban Edinburgh, I choose to journey to natural areas to walk, to rest, and to explore for many reasons. This choice is telling of a disposition—I believe that these excursions are positive, for me. The question of what this kind of activity may offer to others or to groups and of what manifestations within these spaces may tell us of the phenomenological experience of natural spaces is still left pertinent. It is necessary to discover how these kinds of activities are evaluated.

> Open air recreation provides people with great benefits for their health and well-being and contributes to the good of society in many other ways.[1]

The Scottish Outdoors Access Code, quoted above, intimates that the 'right to roam' has been put in place due to the belief that access to these kinds of rural natural and urban green spaces relate unequivocally to human health, wellbeing, and the good of society. In Scotland, then, policymakers and parliament, in the majority, have agreed that this access is positive.

When considering groups of individuals who venture out into these spaces I wondered, in what ways this access could be positive for the individual and the collective? When I began this project, I had several questions in relation to these group activities:

> Why do people journey into natural spaces in groups?
> What do they do within these when they get there?
> And, what happens to people's emotions, physicality, and social dynamics when they do enter these spaces?

To begin to answer these questions, it became clear that the way that people evaluate these activities must be understood. For this reason, we must ask:

> What is positive about this kind of landscape engagement?
> What do activities in natural spaces provide?
> In what ways are group excursions into natural spaces considered to be personally and socially transformative?
> And why?

> In what ways is getting outside good for us as individuals and, as the Scottish Government seems to advocate, good for society?
> And finally, what could be meant by 'good' in this context?

Wellbeing and self-transformation in natural landscapes is concerned with why urban-dwelling people make excursions to natural spaces in Scotland and why people believe that these kinds of activities are positively transformational. It is about the ways in which groups of people consider engagement with natural space as linked to positive mental wellbeing. This book will consider individual, internal and external group motivations for journeying from urban environments in pursuit of natural spaces. This book is interested in what goals certain people within Scotland had in mind and what outcomes were achieved when journeying to these natural spaces. Motivations, as specified by group practitioners were varied: some involved everyday recuperation, some medium-term recovery through short visits, and some involved major attempts at radical lifestyle changes or mental transformation.

Transformative Nature and Scotland

Notions of nature as inherently good for the human condition are not a contemporary phenomenon. It is still however, an area that researchers are aiming to comprehend. The Scottish natural landscape has long been represented as transformative. The natural landscape was considered this way within the philosophy of Scottish common-sense realism in the eighteenth century,[2] influencing transcendentalism around the world.[3,4] It has been celebrated through the 'ancient,' and controversial, Celtic Ossian transcriptions of James Macpherson.[5] It was considered with awe and veneration in the metaphysical aesthetic of Burke's sublime[6] and of the Romantics[7] in the nineteenth century.[8]

Relationships with the natural landscape have continued to be culturally understood as deeply affective, through the writings of Nan Shepherd[9] among the Cairngorms in *The Living Mountain*[10,11] during the Second World War and the work of John Muir in the late nineteenth and early twentieth centuries. Muir firmly placed Scotland's natural landscapes on the world map. He did this by taking his philosophy and conservation of wilderness to America. This was also in relation to the positive worth of nature engagements. More recently, the 'new nature writing' of Robert

Macfarlane,[12] Amy Liptrot,[13] and Kathleen Jamie[14] have further raised interest in Scottish landscapes.

In Scotland, there are longstanding traditions of engagement with the land. These traditions span from the admiration of the Celtic philosophers, folklorists, poets, authors, and artists,[15] as well as beliefs of Scottish travelling communities[16] to industrial-agricultural relationships and contestations of land access and ownership spanning decades.[17] Following this there has also been a resurrection of these traditions within ecological, permaculture, sustainability, and biodiversity endeavours of late, as well as a fresh interest in questions of land ownership in recent years.[18] For some, the Scottish landscape is still a place of agriculture, forestry, and fishing labour,[19] and of course for the Scottish Tourist Board, farming bodies, and to natural material traders, the landscape is an economic commodity. For some this land is a home; for others, a space for extreme sporting activity. More relevant perhaps is the longstanding tradition within Scotland of rambling,[20] camping, and leisure time spent within the landscape.[21]

The Scottish natural landscape, like in many countries, is rooted in culture. One could say that Scottish culture is very much rooted in the natural landscape. This is clear in the prevalence of semi-wilderness images in national iconography, folklore, and history, as well as in the imagery used in Scottish tourist literature. The constructed imagery and iconic landscape scenery is a created or 'invented'[22] idea of Scotland. This is how we have 'seen nature.'[23] For Olwig, 'landscape is often conceived today in scenic, pictorial terms'[24]; however, for the groups with which I explored, the Scottish landscape is not only seen in image. Though 'nature as good' is indeed historically situated, it is also physically and emotionally experienced by the people in these groups. Individuals create their own social relationships within and with the natural landscape. It is not only conceived through historical and political discourse, but also, though influenced by these, experienced and subsumed into the personal narrative of individuals.

Popular culture in Scotland encourages people to engage with the outdoors, whether in taking a walk, cycling, camping, or taking part in general physical activity.[25] For over ten years, every person in Scotland has had the 'right to roam'[26] the outdoor 'natural' landscapes of Scotland.[27]

This means that all mountains, farm land, forested land, rivers, lochs, reservoirs, coastline, margins of cropped fields, grasslands, paths, tracks, and woodlands, day and night, are openly accessible to the public of Scotland.[28] This is regardless of land being privately owned, simply provided that basic rules are abided by.[29] This is unique in the United Kingdom: in Scotland people have more access rights than in England and Wales. In fact, according to The Ramblers Association: 'Scotland has some of the best access rights in the world.'[30] In England, for example, the public have the 'right to access *some* land for walking or certain other leisure activities'; this land must be specifically 'open-access land,' and must not be privately owned unless prior permission is granted.[31] For Vergunst, Scotland's legislative approach that provides access to Scotland's outdoors in 'its entirety' is radical. This allows 'a national sense of an accessible landscape'[32] and group activities in the landscape that are accessible to all.

To say that the Scottish landscape is, and activities within are, accessible to all perhaps assumes that (a) all people in Scotland are aware of their 'right to roam' and (b) that all people in Scotland are able to or do take advantage of this 'right to roam.' I have found this not to be the case with the groups that I have engaged with. To be wholly accessible would presume inclusivity and commonality of access to experience. Often, however, individuals are not aware, nor do they have experience of these landscapes. As Wong states:

> Environmental participation has been dominated by the white middle class in Britain. The analysis is that people whose lives are in order, can devote their energies to be a workforce for nature without asking for anything in return. Access to the enjoyment of open spaces such as National Parks or the open countryside is a normal part of their lives as *they are well informed* and have the resources to go wherever they wish. Whereas, the life situations of disadvantaged groups beset by problems necessitates the combination of being able to see environmental engagement deliver an impact on their lives […] Neither do they have the basis for motivation to care for spaces which they have never had the opportunity to enjoy. *People simply do not know where to go, have no concept of what enjoying the countryside means* and in many deprived areas, there are few if any local greenspaces or pleasant open spaces.[33]

Vergunst[34] comments that the access code is highlighted in tourism offices, outdoor pursuits shops, on billboards, television adverts, and websites. However, this presumes an already interested and engaged party. This promotion of the 'right to roam' does not imply that individuals who are not already engaged would be aware of their rights in this context nor that they would take advantage of these rights. Additionally, those with little personal context in which to place these rights may be unaware of how to utilise them. The Scottish Outdoors Access Code, however, in considering all outdoors space as equal in terms of access rights, 'discouraged some of the traditional mystifications of Scottish landscape.'[35] Demystification of the landscape and a promise of accessibility was considered a move in a positive direction in terms of encouraging engagement with these landscapes. However, what was at stake was

> [T]he confidence with which people in Scotland t[ook] up their new rights [...and] whether the access rights [were to] become embedded within the ordinary ways of moving around the land.[36]

Vergunst also suggested that the ideas of landscape apparent within the code were

> a way into investigating what kinds of relationships between people and their environments [were] being hoped for and [were] actually present as people [sought] access to the outdoors.[37]

It was a recurring statement throughout my fieldwork that people within these groups were not only seeking 'something,' often intangible, they were also finding 'something' deeply intangible and often ineffable within these shared experiences of excursion to natural spaces. Some were seeking social relationships, and some were seeking to try 'something different.' Every individual wanted to achieve 'something' positive by being outdoors, and all were influenced by their own life situations. It is this 'something' that is being explored throughout this book specifically in relation to these groups and these landscapes. In exposing this, we may understand what kinds of relationships are dynamic, what is it that is hoped for, and what do people in fact experience throughout these excursions.

The groups that make up my case studies within this book encourage individuals to behave in a pro-social manner, to move away from their everyday situations, and to place themselves in a scenario with others in the outdoors. With this comes a distinct sense of apprehension. For some this was due to the ambiguity of not knowing what to expect from the excursions. These excursions provided something unfamiliar, but also a promise and an expectation to participants. These journeys brought with them a sense of adventure—or ending up in 'the middle of nowhere' as put by one of my group members. These spaces, though not strictly wild nor even natural (I will come to this point in a moment), brought with them a sense of expanse for the informants, and a distance from the everyday was evident whether this was metaphorical or literal. People wanted to 'get away,' from the urban and their everyday. In a sense, people wished to swap urban culture for a temporal and new cultural landscape with subverted rules, behaviours, and understandings.

Far more than travelling across merely physical landscape boundaries, these journeys provided challenges, new experiences, unfamiliar territories, and new cultures. These journeys and these *other* landscapes allowed for new ways of interacting and new ways of thinking with people and space. They were novelty excursions that allowed for subverted group and individual norms. This was not only with regard to the physical location nor the act of journeying but within the more mundane subversions such as sleeping situations, alternative diets, the sharing of food and housekeeping duties, lack of facilities, lack of access to the internet, and lack of technology—all of which I touch upon in the book as this study has relied solely on ethnographic methods. I went with groups from multiple locations in Edinburgh and Glasgow to many rural natural spaces around the lowlands, highlands, and islands,[38] where I slept alongside, ate alongside, and worked, walked, and played alongside my research groups.

Throughout the book I consider individual and group context as well as associations with one's personal narrative as components of the self that directly relate to not only one's level of activity within natural landscapes but their motivations to partake in activities in such environments and subsequently their experience in these circumstances.

Chapter Outline and Themes

The opening chapter has been devoted to attempting to outline the elusive and ineffable that we will explore here: the phenomenon that is the perception of nature as positively transformational and allied with human wellbeing. Chapter 2 outlines the necessity for a transdisciplinary ethnographic methodology and research strategy to begin to comprehend this. This incorporates how one can adapt performance analysis and concepts of framing to understand these circumstances. It also considers how we might understand experience when abstraction of experience is not always possible for individuals.

The groups of people, and indeed the individuals within these groups, with whom I spent time during my fieldwork were diverse. They had differing mental and physical needs, learning styles, abilities, attitudes, values, socio-economic status, genders, and ages. Each person, of course, had a considerably different personal narrative to the next and with this, differing previous experience of and knowledge of the Scottish natural landscape or indeed natural space. Each of my case study groups is discussed at length in Chap. 3 when we look at 'Getting out, Goethe, and Serendipitous Ethnography'.

This research project was concerned with the social facilitation of groups, including how groups were formed and why, and the agenda and the motivations of groups are also discussed in this chapter. Primarily this research was concerned with the self. It was concerned with the self as part of a group, the perception of the ideal and ought self in relation to motivations to journey, and the self as part of a group within the landscape. The self and group here are dynamic and relational subjects with regard to positive transformation and interaction. This is discussed in Chap. 4 when we approach the journey in relation to belonging and the self. Here, I discuss performed identities, the extended self,[39] and projections of the self in relation to the group, group ethos, and the land. This is considered alongside group dynamics, the development of the group as a process, and collective intention and obligation. This brings discussions and analysis of data with regard to the development of the group as a process, tensions within groups, and plural subjecthood (Gilbert 1990).

In Chap. 5, I discuss these excursions as liminal, yet I do not use the term in the traditional sense. The liminal is a threshold, a point or moment in which a person or a group could be said to have changed due to experiencing something. I describe these points then as transition points and in relation to Turner's concept of bonded, cohesive communitas[40] amongst the group, as brought about by liminal encounter. With this we see the importance of temporality and structure or indeed anti-structure within these excursions as something which aids in the perspective that they are transformative encounters.

Within Chap. 5, I also consider notions of perceived affordance within groups and how this changes throughout experience with the increasing ability to associate ideas and abstract experience within one's personal narrative. By engaging with the group in natural spaces in this manner, we can comprehend notions of place identity. I speak of the sense of self within these experiences as a liminal site. I discuss the group collectively as a site of dynamism and thus liminality before arguing that this then allows for the way that the landscape is perceived to be a site of liminality.

A key realisation within fieldwork and data analysis was that each group differs in how they consider the landscape as something to be utilised, instrumentalised, personified, or anthropomorphised. With this comes a perception of anthropocentric mindsets and the influence of these ways of thinking on experience. This is addressed in Chap. 6. Within my research there are multiple variations of what the social within these excursions entails. For some informants, 'the social' refers to the group and self-dynamic. For others, this dynamic also includes the non-human. This is also discussed in this chapter, alongside the perception of intention and agency within these social networks. Here I look at what is defined as the social group and why this is important when considering the effect of these excursions on mental transformation and sense of self. I will discuss the perception of reciprocity and empathy on one side and altruism, development, and human agency on the other. I will then explore the fine line in between personification, anthropomorphism, and metaphor whilst approaching language used by the informants. Ultimately this will allow for an understanding of the relational subjects within these interactions and the subsequent perception of transformation and what it means to be a good person to groups and individuals in Chap. 7.

The Flow of the Book

The book could be considered as made up of three sections. These sections flow and follow the stages of my research project, as I felt them. The book can be dipped in and out of, though due to the movement from post-field through fieldwork and then analysis; it is written with a journey in mind, and this should be evident throughout. Theoretical underpinnings will be found throughout the first section, methodology throughout the second, and findings and analysis in the concluding section. The chapters provide the twists, turns, and complexities of methodology and analysis. These parts are intended to aid the reader in understanding my research process; however, this ethnographic research may not have transpired in such a linear fashion. This structure is merely navigational, as the reader will see within the chapters. The three distinct phases of pre-, during, and post-fieldwork, as well as the 'what next stage' are often difficult to discern within the research process. These loosely formed sections signify a moving forward whilst the chapters allow us to dwell on ideas that are in between the theoretical strategies, the field, and analysis.

The first section is where the book has been 'set up.' This is where I have introduced the subject, the initial questions, and chapter outline as you have seen. This section is also where I lay out the current and past research trends within the fields of wellbeing and human/nature research. Here is where you will find the project's contributions to these specific fields and more widely. This section also has a focus on the methods and methodology as they were considered pre-fieldwork and moves on to discuss how the field influenced a more responsive and flexible methodology. This includes an outline of how this research project adopted a transdisciplinary research strategy. The second section is devoted to fieldwork. This furthers my argument for a wholly responsive mode of ethnographic research based on learning from the field: this section outlines the serendipitous ethnographic framework. The third section is focused on analysis post-fieldwork. This reveals group and individual motivations. This section also takes analysis further, finding ambiguities regarding the self and liminal threshold as a metaphorical site. It then uncovers further complexities with regard to what is considered social within these encounters and where I lay out my final ideas regarding the field and data.

Before we can move on, I feel it necessary to lay down some definitions that are sometimes contested in this area. The definition of terms detailed here helped me to move forward with my research without remaining stagnant, stuck in overintellectualising or tripping over linguistics. These definitions are nothing more than a clarification allowing us to move forward and into the book together. Some terms are disputed throughout; some remain true to their pragmatic use. These are not considered 'jargon' terms; however, they did pose problems in understanding throughout the research process. By detailing here, I hope to avoid the same sense of stagnancy for readers. I do not, for example, aim to define categorically what the modern world means by 'nature' nor 'transformation' in the wider sense but instead to clarify what is meant by these terms within this book.

The initial terms below are detailed regarding my approach to the field—what was I looking at/doing/working with? 'Natural' landscape was a notion that haunted me throughout this project. Not a philosophical but a practical definition was needed here. I do not mean to imply that the human experience or perception of 'nature' was simple. It is complex; however, in the first instance I needed to clarify, to myself, what landscapes I was speaking of.

Situating 'Nature'

Firstly then, how do I situate nature within this research project? What nature am I speaking of and where? There are two ways to comprehend this:

1. Nature as an area of research—in terms of the physical sciences, history, and human impact upon 'nature' and 'natural' landscapes.
2. In terms of the understandings of the location and presence of and the experience of 'nature' as emergent from the field.

The first mode of thinking acknowledges that the spaces in which I investigated human relationships could most likely no longer be defined as natural. However, the second acknowledges that, within the field, informants considered these spaces as so. Within the book, I focus on the later.

Having fought with the word 'nature,' I finally settled with Brook's use of the term natural landscape, as in

> [T]he way it functions in lay persons' language [...] things like the countryside, beaches, woodlands, national parks, nature reserves, waterfalls and so on.[41]

The nature of which I speak is not what is termed 'nearby nature' by the Kaplans[42] but instead rural 'nature' deliberately, with an objective, travelled to from urban environments. The spaces in which these activities took place would be considered as 'public nature'[43] as opposed to 'private nature.'[44] The landscapes encounters are also not often 'first nature.'[45] The nature experiences of which I speak are not with house plants, images, or videos, as defined as nature contact by Capaldi et al.[46] I consider 'nature' experience as circumstances that involve people within outdoor space and out of urban environments. Brook's use of the term is helpful to adopt as opposed to wrestling with the notion.

The contention over what is meant by 'nature' bears no representation to how the term is used by the groups whom I seek to represent in the field. My research was an enquiry into how these spaces were experienced and perceived rather than how they were linguistically defined. The language used by informants frames these experiences and will be discussed throughout this book.

The amount of previous experience visiting natural spaces amongst my informants varies considerably. There are those who have had not only adequate access to these kinds of spaces since childhood but also a vocalised lifelong interest in and desire to be outdoors. There are also those who have been brought up in what are, considered to be, the most urban of places. Some have spent most of their lives in Scotland's central urban cities such as Edinburgh and Glasgow and all have made a conscious decision to seek out, increase their visits to, appreciation of, and amount of time they spend within 'natural' spaces away from urban areas. Some of my informants have seen green and/or natural spaces rarely and refer to natural spaces as alien or 'not for them.' Experiencing the natural then, refers to excursions into spaces of which, on an everyday basis, individuals did not usually engage. The differences between case study groups in terms of intention to visit these spaces came with a varied set of definitions and associations regarding what kinds of spaces we were visiting. Each of my case studies looks at a different group with its own distinct reasons for visiting natural areas.

Whilst some informants know 'natural' or 'wild' space may no longer be a reality across Scotland, most of my informants are not aware of this. For the majority, living, green space is considered as natural, and accurate definitions do not consciously effect experiences. Of course, distinctions are made between flower beds and meadows or hedges and forests but open-access, public, green, outdoor areas where people may recreate are considered to be natural spaces for the most part. In other words, a hill is a hill, and a river is a river, and a tree in a woodland is exactly that. There are not many further judgements of the space as one thing or another, instead only how it feels to be there. It may feel 'wild,' it may appear 'natural' and whether it is in fact either of these things, when we consider experiential accounts, seems rather beside the point. When talking of shared experience of journeys into the non-urban spaces of Scotland, the debates amongst academics and environmentalists regarding definitions remain, in many cases, irrelevant and to my groups, often unknown.

For Allan, an informant in the Inner Hebrides, the 'academics' who deem his island 'not wild' 'should come and experience it in the winter and see if they may change their minds.' For Allan, the difficulty is in how the landscape is experienced and how we lack the language to attempt to explain these experiences without using these terms that are so contested by the professionals and experts. Allan, perhaps unwittingly, posed a persuasive argument against essentialist definitions of wilderness. Encouraging me instead to have had more interest in the diversity of subjective experiences of these different environments—I wonder, to whom does it matter whether and how we define terms such as 'wild,' 'natural,' and 'remote'? And why does it matter to these people? Perhaps if this book was concerned with ecological conservation or development rather than human experience, this may have been an important issue; however, when writing this book, I was concerned with how several groups of individuals experienced the spaces to which they travelled. Accurate definitions were not of primary concern to these people. I was interested in how these experiences were articulated and how the relational dynamics between the person, group, and landscape affected these experiences.

Realistically we might say that the areas travelled to fall along a spectrum from natural to urban. Spaces visited were canals beneath motorways, forested pathways and farmed hilltops, beachfronts, country estates used for personal and social developmental purposes, industrial outskirts, country manors, islands, low lands, and highlands. The reality is that all the landscapes that were visited were either farmed or surrounded by forestry. The lands were planted, dammed, and fenced. They were also full of living, growing, and flowing things. They were open to the sky, green, brown, dirty, wet, and often cold. For all intents and purposes, these landscapes were natural to the groups without the desire or necessity to question otherwise. These spaces are perceived by participants as natural. For many, perhaps these spaces remain undefined. Participants within excursions did not speak often about farmed land or managed land and certainly not colonised land. They spoke of their experiences within 'nature' and with each other. What is important is not the geographically or topologically accurate definitions of these spaces but the experiences of excursions from the norm, excursions to spaces that may be considered as natural in comparison to the urbanity of these informants' daily lives. The important aspect for my research is that groups of people journey to places such as these with the aim of achieving something, often personal and positively transformational. These people are enthusiasts for what these various spaces may offer and of how shared group experiences and changing personal or physical dynamics may provide something beneficial to the social. They facilitate groups or are participants within groups that journey beyond the urban. I joined them in these natural spaces, to explore why, to what effect, and how these spaces could be considered as transformative.

For this reason, I will refer to natural space to represent that this is how these spaces were so often referred to. The reader can rest assured that I am aware of the truth of the matter: these spaces can no longer be defined as natural regardless of them being experienced as such. I will also use rural natural, always again, referring to the way in which people described these spaces to me.

> [dualism produces] "wilderness" as the place where people are not, as the only place where "true" nature is found, leaving us with "little hope of discovering what an ethical, sustainable, honourable human place in nature might actually look like" [...] Wilderness is a dangerous concept that sets "human beings outside of nature."[47]

Though throughout this book I pay heed only to what is defined as 'nature' or 'natural' by my groups, it is important to acknowledge 'nature' as a contested term. In her undoing of nature (with reference to Judith Butler) throughout her ethnography on Harris in the Outer Hebrides, Mackenzie[48] discusses how nature can be 'recruited in the interest of capital accumulation.' She believes that nature can, and has been reinvented in the interest of capital and that it is important to disturb 'the norm of nature as external to the social' and to 'open up meanings of the terms "nature" and "the social".'[49]

Mackenzie believes that the social is present in these 'natural' landscapes, and it should be represented as so. This idea 'forces us to take responsibility for how this remaking of nature occurs, in whose interest, and with what consequences (for people, plants and animals alike.).'[50] When nature is depicted as not inclusive of humans, and as marred by human inhabitation, we fail to acknowledge the truth—that nature is not separate from society.[51] For Mackenzie, it is this concept that 'underpins processes of privatisation.'[52] By separating communities, the social and culture from 'natural' spaces, the landscape and thus communities within these landscapes are damaged. This silences people and hides the power at play.[53]

I agree with Mackenzie in that definitions of what is and is not natural in terms of landscape is problematic. These definitions tend to further the dualism of nature and culture. Within my case study groups, I approached the social relationships within natural spaces. Regardless of definition, it is imperative to acknowledge the social and cultural within the natural landscape. The social in nature is 'negotiated through the social practices' within and throughout history prior to the Highland clearances and up to date.[54] This is fundamental. For Mackenzie, this is to do with 'collective rights to the land'[55] and for me it is important due to considering access to natural landscapes for wellbeing.

Liminal Sites/Thresholds

Throughout this book, the term liminal threshold will be considered in two different ways. The term first appears in its literal sense: a physical threshold separating the before and after of experience. This is referred to often. We then encounter the liminal threshold as a metaphorical site within the sense of self, the group as a collective, and the perception of the landscape. The liminal threshold or sense of self as liminal site metaphor is useful in envisioning what I believe is going on in these situations. This becomes more complex when we consider consciousness and the location of the mind.

Of course, all thresholds are liminal, and all liminality is about some kind of threshold, whether real or imagined. However, I use deliberate tautology here when I refer to liminal thresholds. The presence of 'liminal' before 'threshold' is connotative of the ambiguity, flux, and change that I wish to suggest here. It is the threshold that is metaphorical. The sense of self is liminal in that it is in flux, ambiguous, and in figurative movement or transformation. It should be considered as a moment of transition.

I consider the notion of transformation as a process. The indicators of positive transformation are somewhat ephemeral and intangible. Indicators are found in the full analysis throughout almost the entire book and are still somewhat elusive. Due to the complexity of each individual's perception of their own self and subtle indicators of transformation, each case must be considered separately. From these individual perspectives, due to analysis, I am able to comment on how these excursions may be considered as transformative beyond any one individual.

Nature and Wellbeing

This book concerns itself with human ecology, the non-human landscape and the synergies, energies, and the cognitive and non-cognitive dynamics within these situations. This places in the spotlight social performance, social roles, and identity theory. This puts the personal, social, and non-human in a relational dynamic. For these reasons, the theoretical framework that has developed begins with general wellbeing and landscape literature and builds into a full secondary research strategy encompassing social identity and group theory, performance, philosophy, social anthropology, psychology and eco-psychology, human geography, sociology, and more. Though prior to fieldwork it was necessary for me to build a coherent picture of what research was available within the remits of wellbeing and experiential outdoor recreation, this is where we will begin.

Due to being primarily concerned with how and why making excursions to natural landscapes may be deemed transformative, particularly positively transformative, this research could be placed somewhere in the remit of wellbeing research. I was aware of this tentative positioning of my work prior to fieldwork, so I began within this domain. The dominant literature with regard to these themes come from within positive psychology and social theory, particularly if we do suppose these endeavours as relating to human wellbeing or flourishing.[56]

Wellbeing, and notions of it, have been addressed extensively by social and political sciences.[57] This area of research gave my own research, though transdisciplinary, a cohesive departure point. This departure point might be considered as an island that prevented drowning in so many disciplinary fields. I was able to return to this 'island' prior to fieldwork and throughout. Wellbeing literature, due to its universal appeal and cross-disciplinary reach provided a platform of which my own work could develop into its own niche. From this point, prior to 'the field,' I moved onto the work that appeared to link the two concepts: 'nature' and wellbeing.[58] It is here that my research may more clearly begin to be contextualised.

> Biophilia has been described as our affective responses to nature and natural environments, each of which has its own "peculiar meaning rooted in the distant genetic past [...] an innate tendency to focus upon life and lifelike forms, and in some instances to affiliate with them emotionally".[59]

Within the remits of human interaction with 'nature,' three primary theories aim to understand why experience within nature may be 'good' for us: (1) the 'biophilia hypothesis,'[60] understanding this need for connection as something innately human; (2) 'attention restoration theory'[61]; and (3) 'stress-reduction theory.'[62] The biophilia hypothesis, as outlined by Hinds and Sparks,[63]proposes an innate human need to 'affiliate with life and life-like processes'[64]; however, advocates of this hypothesis, such as Kellert[65] and Kahn,[66] suggest that this innateness may in fact require additional learning, culture, and experience.[67]

Stress-reduction theory states that

> [E]xposure to certain (unthreatening) natural environments that were evolutionarily beneficial for wellbeing and survival automatically elicit a variety of stress-reducing psychophysiological responses.[68]

And finally, 'attention restoration theory'[69] speaks of a restoration of mental capacities and energy in contact with nature and vegetation, reinforcing 'our spontaneous attention' and allowing our 'sensory apparatus to relax.'[70] The Kaplans believed that there are four factors in this: 'being away,' (as I too suggest when I consider informants leaving urban contexts), 'extent,' 'fascination,' and 'compatibility,'[71] and for the Kaplans a factor that may suggest a preference to one space over another is familiarity.[72] Since the late 1980s and early 1990s, few further influential theories that may explicate the positive benefits of nature experience have emerged.

Later, from the beginning of the twenty-first century to date, much research has claimed that there is a positive effect on social interaction due to nature encounter.[73] Some have found that short times spent within natural environments are seen to improve mood and increase positive emotions.[74] Others claim the benefits of exercise within 'green' environments on mental capacities.[75] Most famously, it is still the Kaplans and their notions of 'attention restoration'[76] that is frequently returned to, to describe the restorative and therapeutic benefits of 'nature' within the literature.

Being outdoors within 'natural' space is good for you.[77] This is the common trope. *Good* may suggest a 'flourishing in life.'[78] These kinds of engagements evoke positive emotions, improve mental health, are restorative of cognitive abilities after mental exertion, and reduce stress.[79] It has been found that, in comparison to built urban environments, exposure to nature can decrease arousal and the perception of stress levels.[80] This is also made explicit in Ulrich's[81] 'psychophysiological stress reduction framework.' Others have shown that, in relation to stress recovery, blood pressure can be decreased by being within natural environments.[82] Additionally, further research claiming that nature protects or at least minimises the impact of stress on mental capabilities has been carried out.[83] This previous research is also supposedly given more public credibility (more credibility is given to this kind of research by funders) presumably due to the fact that it can be medically proven that nature improves physiological health, due to empirical research looking at the heart, brain, and body.[84,85]

Though there has been little else in terms of deciphering exactly *why* excursions to natural landscapes are experientially *good* for individuals and groups, there has been plenty of work arguing that we, in the twenty-first century, have lost our connection with nature. As part of this, there has been an increasing discourse stressing the importance of reconnecting with 'natural' environments. This however is with little evidence, aside from apparent physiological benefits, in terms of explicit reasoning as to why in fact we should bother.

Connection[86] and Disconnection

Questions regarding human connectedness to nature within the western world have been broached across the social sciences, cognitive sciences, and medical sciences with increasing interest, particularly over the past two decades. This subject has gained in momentum rapidly since the beginning of the century and to date.[87] Perhaps the most recognised term for a lack of connection to nature is Louv's[88] 'nature-deficit disorder,' a pseudoscientific term coined to refer to the contemporary generation's (Generation X and the millennials[89]) lack of contact with the natural world and the concern that this causes.[90] Louv is far from alone in this concern:

> People are not as connected to nature as they could be and this has implications, not only for the wellbeing of the environment, but also for the well-

> being of individuals. In fact, there is growing evidence that supports the age-old belief that connecting with nature promotes flourishing (i.e., enhanced hedonic and eudaimonic wellbeing) and positive mental health.[91]

Varied disciplinary approaches to nature connectedness however seem to come to the similar conclusion that a connection to 'nature' is something that is important, valuable, and positive to human wellbeing. Yet, little sufficiently accounts for why within the social sciences, particularly qualitatively. Capaldi et al. have found a significant relationship between this and 'happiness indicators such as positive affect and life satisfaction.'[92] They do not however offer any further experiential evidence nor explanation. Nature connectedness is also linked to psychological resilience, that is the ability to manage stress and maintain 'positive mental health'[93] which again brings us back to what we already knew with regard to the 'stress reduction theory'; for some, being outdoors reduces stress levels. However, this is not the whole picture.

Mayer et al.[94] found that people asked to walk within natural surroundings spoke of reduced public self-awareness, and there have been close links here with time spent in nature and increased self-autonomy. The idea that exposure to nature is *good* for you is certainly not a new phenomenon. It is a recurring theme throughout history.[95] This of course though is understood as an individual difference,[96] meaning that positive feelings within natural space are subjective and depend invariably on 'one's subjective sense of connection to the natural world.'[97] This is also believed to be the case by Mayer and Frantz[98] who state that those with a higher self-awareness of being connected to nature will spend more time outdoors.[99] Mayer and Frantz[100] indeed went to lengths to create the Connectedness to Nature Scale (CNS), allowing them to measure levels of emotions of being *connected* with the natural world. This too lacked the qualitative experiential accounts that seem so key in comprehending the ineffable and intangible connection or disconnection that is felt culturally.

Nisbet, Zelenski, and Murphy, a little later, in 2009 (still only less than ten years ago) proposed something else: a different construct named 'Nature Relatedness.' This also aimed to describe a connectedness to, appreciation of and understanding of nature as well as environmental behavioural values (an example of the instrumentalising of positive emotions discussed later). Later, in 2011, they used this scale to decipher that positive effect, autonomy, and personal growth were positively associated with, what they call, 'Nature Relatedness.'[101] Whilst we were now able to decipher to what extent people felt connected to nature thanks to Mayer and Frantz,[102] we were still unsure of why and what this meant to individuals.

Thanks to Nisbet et al.[103] we could see that people felt a positive effect—they felt a sense of autonomy and personal growth due to being within nature; again, we still did not know why. Similarly, Zhang et al.[104] found that those who were more 'emotionally attuned to natural beauty' gained the most in engaging with 'nature.'[105] What this meant was unclear in the sense that gaining 'the most' cannot be objectively measured nor were the positive experiential benefits of being attuned to natural beauty explained qualitatively or subjectively.

It seems highly evident that qualitative accounts are necessary. Experiential accounts coming from people with varying social contexts and narratives would clarify this perception of personal connectedness to 'nature.' One might wonder whether the above rhetoric is wholly inclusive: do these theorists and researchers contemplate the implications of social or socio-economic backgrounds? It is perhaps an obvious point to make, however, that those who enjoy being within 'natural' environments are more likely 'to be motivated to engage with it more frequently.'[106] Not only this, but those who have had more experience within natural space 'tend to experience more positive experiential states.'[107] Hinds and Sparks[108] found that those who had grown up in natural locations, in comparison to those who grow up in urban locations did report 'more positive affective wellbeing and stronger environmental identities.'[109] They also state that physical demographic variables have no inference on the sense of psychological restoration.[110] In Edinburgh and Glasgow, physical demographic variables such as distance from green space often have a lot to do with socio-economic status, levels of education, and previous experience of the outdoors. This means that for some there is a lack of personal association with natural space—this influencing not only the way that these environments are perceived but also how they are experienced. In terms of experience, as others haven't, I explore this further.

Brook[111] claims people may recover from illness at a faster rate should they be exposed to natural elements, particularly nearby. In line with this, she too suggests that those who do not have nearby nature are more likely, statistically, to suffer with medical and mental health problems.[112] She suggests reasons as to why this is so. According to Brook,[113] and the numerous studies she refers to regarding children and nature and the lack of connection that many are said to have to the natural world, this is detrimental to the development of motor skills, balance, and creative and social and emotional skills, as well as 'spiritual fulfilment.'[114] This subsequently is said to have a detrimental effect on life choices made, relationships, and prospects.[115] Once again she states that for a 'lay person looking at the evidence [this] seems straightforward.'[116] She also, as many have before her, suggests that a lack of connection, due perhaps to 'limits imposed on play spaces by parental anxieties and the role of myth and media sensationalism'[117] is a problem.

Though Brook suggests why this lack of connection may have consequences for children, presumably in developmental stages, she does not infer as to the consequences for adults: those who may or may not have had childhood experiences within nature as adults. She also does not comment on those adults who do indeed actively connect to nature nor explain why this may be a positive thing to do. These kinds of research do not address the self, the group, and the landscape as relational counterparts within these excursions.

These researchers believe that experiences within nature will increase feelings of being connected if only momentarily.[118] Some believe that engagement within these kinds of spaces as opposed to merely passive observation of these landscapes 'may awaken the dormant genetic, psychological traits that enabled our ancestors to survive in the natural environment.'[119] These researchers along with the previously mentioned do not account for the phenomenological experience of being within natural landscapes or natural spaces nor do they account for the relations between the self and non-human that are often ineffable. They do not explain fully why being in nature is experienced as positively transformative.

Whilst Louv mourns this loss of connection for the sake of children and Brook for the detrimental effect this will have on 'individual health and community cohesion.'[120] Adams[121] speaks of a detriment to not only humanity but to the natural environment itself. This he calls an 'impoverishment.' He insists that the relationship between humans and the 'rest of nature' defines us as a species. He claims that this interrelationship 'has always been an essential constituent of what it is to be human.'[122] Not only this, he goes on to say:

> this crucial partnership is lived mostly unconsciously [...] we are suffering from an unprecedented and perilous estrangement of a single participant – we human beings – from the rest of the shared earth community [...] Healing this dreadful dissociation is one of the most urgent responsibilities of contemporary life.[123]

A position such as this is not uncommon within eco-philosophy or eco-psychology. It also finds its place in discussions of environmental behaviour. Adams[124] uses such drastic language: 'suffering,' 'perilous,' 'dreadful,' 'urgent,' a common rhetoric too within certain circles of scholars looking at the anthropocene—in what way, could such a drastic positioning be helpful when discussing human wellbeing and its relationship to the natural?

Instrumentalising Wellbeing

> With humans against the rest of nature, we find another tragic version of this destructive dynamic [...] When this defensive reaction is combined with diminished involvement with nature, we have a treacherous situation. [...] If, like Narcissus, we are (pseudo)content to see only our own egoic face in the face of the earth, we will look superficially and find only what we already believe or wish: namely, our own desires and agenda.[125]

Are people's aims for a sense of wellbeing appropriated in a bid to encourage pro-environmental care? Is it helpful or appropriate, when considering mental wellbeing to correlate these two things? When the link is made (rather commonly) between doing things in nature for wellbeing and the push towards doing *our bit* for the planet, do we do a disservice to those who aim for bettering their mental health?

> When our relations with the rest of nature are determined by our ego-centered subjectivity and world-view, we abdicate the fullness of our being and tend to desecrate the natural world.[126]

If we continue here with more catastrophising language from Adams, we see a clear concern with pro-environmental care. For Adams, this should be the motivation to connect with nature. Rather than encounters with nature being pushed for the sake of our own wellbeing, instead it is encouraged as reciprocal endeavour in which man might correct his destructive ways. There is a place for this if concerned with the protection and conservation of nature. However, this rhetoric does not aid in understanding human experiences within nature. This rhetoric certainly does not aid in the understanding of these experiences within a remit in which case study groups engage with natural space for personally positive reasons. Within this remit individuals are, though a part of wider mankind, not intending further destruction of the landscape. For the most part, they simply wish to feel *something*. Adams is not alone in this call for a win-win scenario due to social engagement with 'nature.'[127] Similarly to Brook[128] I agree that a 'continual focus on the environment "in crisis" might not always be the most helpful.'[129]

Unhelpful and accusatory accounts of man, deeming us ego-centred and destructive[130] and outlining the lack of improvement to environmental state and attitudes[131] do not represent the experiences of those who do choose to journey into natural environments for the sake of positive personal or group transformation. For Hinds and Sparks[132] indeed experiences within nature do have correlations with pro-environmental behaviour.[133] It seems that this rhetoric is disinterested in the experience of man in nature unless this experience can be utilised to forge a pro-environmental attitude which for many is not their aim when engaging in journeys to natural landscapes. For many, these excursions are about their own mental wellbeing and personal development whether this is the explicit agenda of groups or not.

The Nature Experience: Sociality, Place, and the Self

As I have shown, there is considerable literature and empirical research within the remits of nature and wellbeing. However, the majority of empirical studies provide quantitative accounts.[134] This book, however, is concerned with the qualitative experience of individuals and groups who journey into these natural landscapes. I am interested in the experiential accounts of encounters within these spaces. Not only this but I am interested in how these encounters are considered as transformative in relation to wellbeing, going beyond the merely stress-relieving or attention restoring, and probing the concept of the innateness of which Wilson speaks within his biophilia hypothesis.[135]

To fully comprehend what happens to people when they journey as part of a group to natural spaces one must begin with individual experiential accounts. One must witness these experiences and observe how people behave within these scenarios. It is necessary to understand the way that, relationally, the self, group, and landscape seem to affect one another. It is necessary to observe how the group dynamic develops and how not only people perceive the landscapes but how they think about themselves and their own experiences within them.

According to Abrams and Bron,[136] when an individual considers aspects of the self they will consider standards derived from their own self-definition, for example, attitudes, commitments, feelings, morals, and needs. This then may help in identifying both those with seemingly similar self-definitions and with whom they may interact—and subsequently activities in which they may take part. When an individual considers themselves as a public entity they will consider standards relating to social norms.[137] When standards relating to social norms are considered this would presumably be within one's own concept of normative behaviour. Accordingly, people consider their social self in relation to others. One could argue then that the choice to spend time with or in doing certain things is implicated in one's self-concept. I will explore this further when I suggest that in joining a group to take part in these excursions one aims to self-identify.

> Identity is a complex topic. At its core, it refers to some way of describing or conceptualizing the self, which may incorporate personal roles and attributes, membership in social groups or categories, and connections to geographical locations [...] Other important attributes of identity are that it is fluid, multidimensional, and socially relevant.[138]

For Devine-Wright and Clayton[139] one's identity is a key mediator in behaviour.[140] It is also fluid, complex, and social,[141] and the physical environment has strong connections with one's sense of self.[142] One's identity then can be linked closely to place, as we have seen above regarding connectedness to nature, but also can be seen in terms such as place identity,[143] place attachment,[144] and sense of place.[145] Similarly, according to Carter[146] when one's identity is verified, individuals 'will experience positive emotions,'[147] the environment will shape subjective definitions of the self and interactions with other people.[148] More importantly, to research findings Carter states:

> Motivated by the desire to feel good, individuals seek to verify their identities among others in their environment.[149]

Devine-Wright and Clayton also support this by stating that as individuals we are more concerned with events that we deem self-relevant, and we behave in order to present a desired view of ourselves.[150] Abrams and Bron[151] state that behaviour within a group is highly regulated by the self-concept[152] and for Goldenberg and Saguy[153] 'group-based emotions'[154] play in shaping intergroup processes.'[155] This is not to say that the group experiences the same emotion as a collective but instead that each individual's emotions may affect and be affected by the group. Ultimately the group, made up of individuals, is emotionally dynamic. Diener[156] claims that within a group context, individuals may lose any sense of self-consciousness.[157] However, I have found this is not necessarily the case, certainly not upon initial meeting, most often before the group engages in any activity. 'Anonymity, lack of identifiability and externally focused attention' aid in what Abrams and Bron refer to as 'deindividuation.'[158] This concept involves a 'sense of unity with the group' as well as loss of inhibitions, and counter normative behaviour[159]

This potentially has similarities to subverted norms and a sense of communitas as Turner[160] defines it. Interestingly Abrams and Bron also state that deindividuation can lead to a loss of sense of self.[161]

If we want to explore the self within these situations, we need to explicitly look at identity. We need to consider how individuals think of themselves within these encounters. It will be within this that understandings of both behaviour and experience will begin. If one's identity encompasses social roles and geographical ties, it must be the case that one's sense of self in relation to these natural spaces will come into play and may become fluid within these spaces and as the group dynamic shifts. This is relevant due to the fact that the intention of this research was to explore social relationships within these scenarios. Importantly this research questioned what is deemed potential social agent within these encounters. This aspect of my research is a key unique contribution to the remits of 'nature' and wellbeing. This research has been an explicit attempt to understand, relationally what is going on between the self, group, and non-human other and to explore how people perceive the social and the relationships within these encounters. It was also an explicit aim to comprehend why and how people then grant these encounters transformational status.

> We can think of self-identity as a reflexiveness that results in the recognition of one's personhood—an understanding of who one is with reference to biography and situation.[162]

For me, the experience of the natural landscape is fundamentally understood in relation to individuals and to how they share in these experiences during encounter and post-encounter. What the previous research that I have outlined fails to comprehend with regard to 'nature' experience is a person's self-identity. Without paying heed to this, how can a researcher comprehend someone's experience? They might be able to decipher whether someone says that he or she enjoys the outdoors or not, they might be able to state that someone feels more relaxed or that their stress levels are reduced, and perhaps they may be able to link this to physiological factors. Yet, they will not be able to decipher exactly what or why someone is experiencing what they are within these groups on a qualitative level: indeed, the intangible, phenomenological or ineffable, within these experiences. Researchers can do little more than say whether being outdoors is *good* for individuals, or not if the case may be, without using ethnographic or at least qualitative methods to understand why these activities are deemed to be transformational.

Notes

1. The Scottish Outdoors Access Code (2005: 10).
2. See Stewart (1829) and Reid (1764/1997) regarding Scottish common sense realism as celebratory of human efficacy, self-reliance, and new insights.
3. Transcendentalists appreciated the innate goodness of nature and its relation to human nature.
4. See Miller (1950), Capper and Wright (1999), and Packer (1995).
5. See Trevor-Roper (2014). The mythical bard, Ossian, represented these landscapes as wild and dramatic. Also, see Olwig (2002) who discusses the sublime and it being 'the semblance of a nostalgic return to nature' and Ossian as the voice of 'a raw and wild nature [...] unmediated by culture' (Ibid: 162). In thoughts on the sublime the landscape was met with intangible awe and veneration.
6. See Hope-Nicholson (1959) and Brady (2013).
7. See Brown (2001).
8. Within the literature of Robert Burns and Sir Walter Scott alongside the likes Of Coleridge and Wordsworth in England.
9. Shepherd (1977/2011).
10. *The Living Mountain* is a celebration of the human spirit amongst 'nature' alongside the mountain landscape in the (now) Cairngorms National Park.
11. See also Nan Shepherd (2014) *In the Cairngorms.*
12. Robert Macfarlane (2003, 2007, 2013, 2016).
13. Amy Liptrot (2015).
14. Kathleen Jamie (2015).
15. For more on literature/folklore and the landscape, see Dorson (1968): literature throughout history arguably now seeps into the readings of the landscape as it is now; an amalgamation of industrialisation, romanticism, mysticism, nostalgia, and misrepresentation. In Scottish folk

memory, there is a palpable affinity or empathy with the landscape. See Burns (2017) for popular examples right through to many more modern works from Cambridge (1999/2008), and many more at The Scottish Poetry Library (2017). See also Crumley (2000, 2003, 2007), Macfarlane (2007), and MacCaskil (1999), and for more on representation of Scotland in film, see Fowler and Helfield (2006). Again, just a small selection.

16. Additionally, there are certainly traditions of 'placelore' within the Scottish traveller community. According to Reith, these 'interaction[s] with oral traditions [and] memories function on many experiential levels to build creativity' (Reith 2008: 77). With Scottish travelling peoples maintaining the 'richest oral culture in Europe,' they often utilised 'narratives that tie[d] together geography and metaphor to recover ancestral memory' (Ibid: 81). A heritage that held a 'holistic view of the landscape' (Ibid: 82): natural, material, ancestral, ethereal, eternal, spiritual.
17. See McVean (1969) and Hunter (1974).
18. See Mackenzie (2003, 2006), Cameron (2010), and Vergunst (2013).
19. The number of people working on the land in agriculture, forestry, and the fishing sector, for example, has declined significantly due to the growth of industry post the Industrial Revolution. Clearly now in Scotland, there are more machines engaged in land work than people.
20. Rambling refers to the act of walking in the countryside for pleasure, in groups normally. Of course, it may also conjure images of simply wandering from place to place.
21. See Lorimer and Lund (2003), and for an idea of rambling in Scotland, see Ramblers (2017).
22. See Trevor-Roper (2014) and Olwig (2002).
23. Olwig (2002).
24. Olwig (Ibid: xxvii).
25. See: The Scottish Outdoors Access Code (2005), The Scottish Waterways Trust (2016), Scotland Outdoors (2017), The List (2009), and Scottish Natural Heritage (2017). This is just a fraction of what is available online regarding the encouragement to get outdoors in Scotland.
26. The Right to Roam/ Freedom to roam, officially known as The Land Reform (Scotland) Act 2003: 'Part 1 of the Land Reform (Scotland) Act 2003 gives everyone statutory access rights to most land and inland water. People only have these rights if they exercise them responsibly by respecting people's privacy, safety and livelihoods, and Scotland's environment' (Scottish Outdoors Access Code 2005: 10).

27. Scottish Natural Heritage Organisation in Vergunst (2013: 124).
28. Scottish Outdoors Access Code (2005).
29. The basic rules or 'The Code' ... 'Exercising access rights responsibly [...] take responsibility for your own actions [...] respect people's privacy [...] help land managers and others to work safely and effectively [...] care for your environment [...] keep your dog under control [...] take extra care if you are organising a group [...]' (Scottish Outdoors Access Code 2005: 05) as well as 'Respect the interests of other people [...]' (Ibid: 12).
30. Ramblers: at the heart of walking (2017).
31. Gov.uk (2017).
32. Vergunst (2013: 121).
33. Wong (2007: 43) emphasis my own.
34. Vergunst (2013).
35. Nadel Klein in Vergunst (2013: 126).
36. Vergunst (2013: 127).
37. Vergunst (Ibid: 125).
38. The journeys to new spaces and the experience of this, were vital in attending my research agenda. This was not necessarily 'multi-sited ethnography' as defined by Marcus (1995): where ethnographies are seen to be 'cross cut[ting] dichotomies such as the "local" and the "global," the "lifeworld" and the "system" [... and are] both in and out of the world system' (Ibid: 95). This research does however sway considerably from the traditional single site field location as Marcus suggests, having not only multiple locations but multiple case study groups. This ethnography did indeed engage with multiple sites across Scotland. It is, therefore, multi-sited in the sense that the research sites were plural. They did however remain within a geographically limited area, and in some regards 'local.'
39. Belk (1988).
40. Turner (1969, 1970, 1974).
41. Brook (2010: 298).
42. in Brook (2010: 298).
43. Brook (2010: 298).
44. see Lokhorst et al. (2014).
45. Conradson (2005).
46. Capaldi et al. (2015).
47. Cronon, in Mackenzie (2013: 27).

48. Mackenzie (2013: 26).
49. Mackenzie (2013: 28).
50. Braun in Mackenzie (2013: 28).
51. Mackenzie (2013: 26).
52. Kats in Mackenzie (2013: 27).
53. Mackenzie (2013).
54. Mackenzie (2013: 27).
55. Mackenzie (2013: 40).
56. See Seligman (2011).
57. See Jimenez (2008), Dolan (2014), Bormans (2012), Zevnik (2014), within social anthropology (see Thin 2005), happiness movements (see the likes of Action for Happiness 2017) and The Network of Wellbeing (2017) by social policy scholars (see Chopel 2012; Frey 2012; Chanfreau et al. 2008) and economists (see Layard 2011; Dluhosch and Horgos 2013; Graham 2012; Kahneman and Krueger 2006); and of course within the remit of subjective wellbeing and positive psychology (for examples of work on Subjective Wellbeing see Diener (1984), Diener et al. (1985), Pavot et al. (1990), Pontin et al. (2013), Tinkler and Hicks (2011), Samman (2007), Kashdan (2004), Hallerod and Seldon (2013), Hills and Argyle (2002), Francis (1998), and the OECD (2013)).
58. Research into the benefits and impact of engagement with 'natural' environments on notions of wellbeing, both social and personal as well as physiological has been carried out worldwide, an amount too vast for detail here; however, for a flavour, see work carried out in the United States (Payton et al. 2008), in Japan (see work on Shinrin-Yoku; Park et al. 2011; Morita et al. 2007), in Canada (Howell et al. 2011), and in mainland Europe (see Lopes and Cananho 2013; Hunter 2003; Esposito et al. 2013). This work, though further afield, is still relevant to the positioning of this book.
59. Wilson (1984) in Hinds and Sparks (2011: 452).
60. Wilson (1984).
61. Kaplan and Kaplan (1989).
62. Ulrich et al. (1991).
63. Hinds and Sparks (2011).
64. Hinds and Sparks (2011: 452).
65. Kellert (2002).
66. Kahn (1997).

67. Hinds and Sparks (2011: 110).
68. Capaldi et al. (2015: 02).
69. Kaplan and Kaplan (1989).
70. in Morris (2003: 13).
71. Kaplan and Kaplan (1989).
72. Kaplan and Kaplin (1989).
73. Groenewegen, van den Berg, de Vries, and Verheij (2006).
74. See Berman et al. (2012), Mayer et al. (2009) and Nisbet and Zelenski (2011).
75. Thompson-Coon et al. (2011).
76. Kaplan and Kaplan (1989).
77. General claims to the benefits of nature on the economy, society, the environment and health and wellbeing are made by Bird (RSPB and 2007), Hartig et al. (2007), Maas et al. (2006), Ward-Thompson (2007, 2008, 2010, 2011), Ward-Thompson and Travlou (ed. 2007) Morris (2003), Steptoe and Butler (1996) Gordon and Grant (1997), Calfas and Taylor (1994), Burns (1998), Ulrich and Parsons (1992), Gruber (1986) and Freeman (1984).
78. Capaldi et al. (2015: 01).
79. See Abraham, Sommerhalder, and Abel (2010), Groenewegen et al. (2006) and Hinds and Sparks (2011).
80. Ulrich (1979, 1981) and Ulrich et al. (1991).
81. Ulrich (1983).
82. Hartig et al. (2003).
83. Stigsdotter et al. (2010), Van Herzele and de Vries (2012), Van den Berg, Maas, Verheij, and Groenewegen (2010) and Ulrich (1981).
84. See Tsunetsugu, Park, and Miyazaki (2010), Moore (1981), Parsons (1991), Ulrich et al. (1991) and Ulrich and Parsons (1992).
85. In relation to physical health, fitness, and physiological benefits, there are many further examples of research, see Bell et al. (2015) and (Brook 2010). For further insights into the body in 'nature,' lifestyle, and tourism, see Little (2012, 2015) and for an alternative use of the term wellbeing in terms of physical health. For an extensive early literature review (beginning in the 1970s through to the early 1990s), particularly in relation to physiological benefits, outdoor exercise, and policy and organisational intervention in Scotland, see Morris (2003).
86. 'Nature connectedness [...] refers to one's subjective sense of connection with the natural world' (Capaldi et al. 2015: 02).

87. This is also something addressed in a number of popular culture non-fiction, memoirs, and autobiographical writing, some of these texts having become award-winning best sellers within the past decade; see Robert MacFarlane, *Mountains of the Mind (2003)*, *Landmarks (2016)*, *The Wild Places (2007)*, and *The Old Ways (2013)*; Helen MacDonald, *H is for Hawk* (2014); and Amy Liptrot, *The Outrun* (2015)—McDonald's text telling the story of recovery from the loss of her father and Liptrot's text being specifically about returning from London to her childhood home in Orkney and recovering from mental illness and alcoholism. The popularity of texts such as these highlights an increasing awareness within popular culture in Scotland and the UK. Similarly, Richard Mabey, writer and broadcaster, has published several books exploring nature and culture (see Mabey 1980, 2005, 2010).
88. Louv (2005).
89. Generation X: born 1966–1976 and coming of age 1988–1994, Generation Y (millennials): born 1977–1994 and coming of age 1998–2006, and Generation Z (also millennials): born 1995–2012 and coming of age 2013–2020.
90. See also Adams (2009) and Brook (2010) and for a mourning for connection to nature see Capaldi et al. (2015) and Hinds and Sparks (2007). For this concept specifically in relation to children, see Brook (2010), Ward-Thompson (2007, 2008), and in relation to a perceived vulnerability that this leaves us with, see Jordan (2009) and for the perception of a desire for reconnection seen in transformative nature natural tourism, see Little (2012).
91. Capaldi et al. (2015: 02).
92. Capaldi, Dopko, and Zelenski (2014).
93. Ingulli and Lindbloom (2013).
94. Mayer et al. (2009).
95. Selhub and Logan in Capaldi et al. (2015).
96. Tam (2013).
97. Capaldi et al. (2015: 02).
98. Mayer and Frantz (2004).
99. See also Nisbet, Zelenski, and Murphy (2009) and Tam (2013).
100. Mayer and Frantz (2004).
101. In Kamitsis and Francis (2013).
102. Mayer and Frantz (2004).
103. Nisbet et al. (2011).

104. Zhang et al. (2014).
105. Zhang et al. (2014: 60).
106. Hinds and Sparks (2011: 463).
107. Hinds and Sparks (2011: 463).
108. Hinds and Sparks (2007).
109. in Hinds and Sparks (2011: 455).
110. Hinds and Sparks (2011: 455).
111. Brook (2010).
112. Brook (2010).
113. Brook (2010).
114. Brook (2010: 305).
115. Brook (Ibid.).
116. Brook (Ibid: 299).
117. Brook (Ibid: 58).
118. See also, Mayer et al. (2009), Nisbet (2013) and Nisbet and Zelenski (2011).
119. Dubosm in Hinds and Sparks (2011: 453).
120. Brook (2010: 307).
121. Adams (2009).
122. Adams (Ibid: 38).
123. Adams (2009: 38–39).
124. Adams (Ibid).
125. Adams (2009: 47–48).
126. Adams (2009: 43).
127. For studies in connection to pro-environmental care and nature protective behaviour, see Stedman (2002), Hinds and Sparks (2007), Sparks et al. (2014), and Stern et al. (2011). In relation to pro-environmental care, place, and identity, see Devine-wright and Clayton (2010). In relation to self and collective efficacy, see Jugert et al. (2013); to destructive behaviour, see Adams (2009); to emotional affinity and caring behaviours, see Kals et al. (1999); to anthropomorphism and its influence on pro-environmental care, see Tam (2013); to the link between witnessing beauty in nature and pro-environmental care, see Zhang, Howell, and Iyer (2014); a love for nature and pro-environmental care, see Jordan (2009); nature considered as a metaphorical 'home,' see Hung (2010); and the environmental benefits of green space as well as green space quality and effects of public health, see Morris (2003).
128. Brook (2010).

129. Brook (Ibid: 309).
130. See Adams (2009) and Shultz et al. (2004).
131. Capaldi et al. (2015).
132. Hinds and Sparks (2007).
133. Also found in empirical studies by Nord, Luloff, and Bridger (1998) and Teisl and O'brien (2003) (in Hinds and Sparks 2007).
134. In relation to environmental connection and identity, see Hinds and Sparks (2007, 2011); in relation to protective behaviours and connection, see Shultz et al. (2004); to health and salutogenic encounter, see Bell et al. (2015); to spirituality and wellbeing, see Kamitsis and Francis (2013); and with regard to the connectedness to nature scale, see Mayer and Frantz (2004). In relation to connectedness to nature and pro-environmental behaviour, see Sparks et al. (2014); to happiness and nature relatedness, see Zelenski and Nisbet (2012), and in relation to public engagement with green space, see Ward-Thompson (2007, 2008, 2010, 2011).
135. Wilson (1984).
136. Abrams and Bron (1989; Ibid: 312).
137. Abrams and Bron (1989).
138. Devine-Wright and Clayton (2010: 267).
139. Devine-Wright and Clayton (2010).
140. See also Carter (2013) who claims that identities are 'motivational when activated in a social situation' (Ibid: 205).
141. Devine Wright and Clayton (2010).
142. Devine Wright and Clayton (2010).
143. Proshanky et al. (1983).
144. Altman and Low (1992), Milligan (1998) and Kyle et al. (2014).
145. Tuan (1979) and Hay (1998a, b).
146. Carter (2013).
147. Devine-Wright and Clayton (2010: 205).
148. Carter (2013).
149. Carter (2013: 205).
150. Devine-Wright and Clayton (2010).
151. Abrams and Bron (1989).
152. Abrams and Bron (Ibid: 311).
153. Goldenberg and Saguy (2014).
154. 'The term group-based emotion refers to emotions that are dependent on an individual's membership in a particular social group and occur in

response to events that have perceived relevance for the group as a whole.' (Goldenberg and Saguy 2014: 581). Additionally, 'collective emotions differ from group-based emotions, because group-based emotions consider an individual's emotional experience in response to group related events, whereas collective emotions refer to the collective as the entity that experiences the emotion' (Goldenberg and Saguy 2014: 582).

155. Goldenberg and Saguy (Ibid: 581).
156. Diener (1984).
157. in Abrams and Bron (1989).
158. Abrams and Bron (Ibid: 311).
159. Abrams and Bron (1989).
160. Turner (1967, 1969, 1974).
161. Abrams and Bron (1989).
162. Cerulo (2009: 532).

References

Abraham, A., Sommerhalder, K., & Abel, T. (2010). Landscape and Well-Being: A Scoping Study on the Health-Promoting Impact of Outdoor Environments. *International Journal of Public Health, 55*, 59–69.

Abrams, D., & Bron, R. (1989). Self-Consciousness and Social Identity: Self-Regulation as a Group Member. *Social Psychology Quarterly 52*(4), 311–318. American Sociologist Association. Stable URL: http://www.jstor.org/stable/2786994

Action for Happiness. (2017). Available at http://www.actionforhappiness.org/. Accessed 17 July 2017.

Adams, W. W. (2009). Basho's Therapy for Narcissus: Nature as Intimate Other and Transpersonal Self. *Journal of Humanistic Psychology, 50*(1), 38–64. https://doi.org/10.1177/0022167809338316.

Altman, I., & Low, S. M. (1992). *Place Attachment: A Conceptual Inquiry*. London: Springer.

Belk, R. W. (1988). Possessions and the Extended Self. *Journal of Consumer Research, 15*(2), 139–168. Oxford University Press. Stable URL: http://www.jstor.org/stable/2489522

Bell, S. L., Phoenix, C., Lovell, R., & Wheeler, B. W. (2015). Seeking Everyday Wellbeing: The Coast as Therapeutic Landscape. *Social Science and Medicine, 142*, 56–67.

Berman, M. G., Kross, E., Krpan, K. M., Askren, M. K., Burson, A., Deldin, P. J., Kaplan, S., Sherdell, L., Gotlib, I. H., & Jonides, J. (2012). Interaction with Nature Improves Cognition and Affect for Individuals with Depression. *Journal of Affective Disorders, 140*(3), 300–305. https://doi.org/10.1016/j.jad.2012.03.012.

Bird, W. (2007). Natural Thinking: Investigating the Links Between the Natural Environment, Biodiversity and Mental Health. RSPB. Available at http://www.rspb.org.uk/policy/health. Accessed 27 July 2017.

Bormans, L. (Ed.). (2012). *The World Book of Happiness*. London: Marshall Cavendish International.

Brady, E. (2013). *The Sublime in Modern Philosophy: Aesthetics, Ethics and Nature*. New York: Cambridge University Press.

Brook, I. (2010). The Importance of Nature, Green Spaces, and Gardens in Human Wellbeing. *Ethics, Place and Environment: Journal of Philosophy and Geography, 13*(3). https://doi.org/10.1080/1366879x.2010.522046.

Brown, D. B. (2001). *Romanticism*. London/New York: Phaidon.

Burns, G. W. (1998). *Nature-Guided Therapy: Brief Integrative Strategies for Health and Well-Being*. London: Taylor and Francis.

Burns, R. (2017, February 23). In A. Turnbull (Ed.), *Robert Burns – Nature: 12 Works Inspired by Nature*. Independently Published. ISBN-10: 1520675674.

Calfas, K. J., & Taylor, C. (1994). Effects of Physical Activity on Psychological Variables in Adolescents. *Pediatric Exercise Science, 6*, 406–423.

Cambridge, G. (1999/2008) *'Nothing but Heather': Scottish Nature Poems, Photographs and Prose*. Bodmin/Kings Lynn: MPG Books Group

Cameron, E. (2010). 'Unfinished Business': The Land Question and the Scottish Parliament. *Contemporary British History, 15*(1). https://doi.org/10.1080/713999393. 2001.

Capaldi, C. A., Dopko, R. L., & Zelenski, J. M. (2014). The Relationship Between Nature Connectedness and Happiness: A Meta-analysis. *Frontiers in Psychology, 5*, 1–1. https://doi.org/10.3389/fpsyg.2014.00976.

Capaldi, C. A., Passmore, H., Nisbet, E. K., Zelenski, J. M., & Dopko, R. L. (2015). Flourishing in Nature: A Review of the Benefits of Connecting with Nature and Its Application as a Wellbeing Intervention. *International Journal of Wellbeing, 5*(4), 1–16. https://doi.org/10.5502/ijw.v5i4.449.

Capper, C., & Wright, C. E. (Eds.). (1999). *Transient and Permanent: The Transcendentalist Movement and Its Contexts*. Boston: Massachusetts Historical Society.

Carter, M. J. (2013). Advancing Identity Theory: Examining the Relationship Between Activated Identities and Behaviour in Different Contexts. *Social Psychology Quarterly, 76*(3), 203–223. American Sociological Association. Stable URL: http://www.jstor.org/stable/43186697
Cerulo, K. A. (2009). Nonhumans in Social Interaction. *The Annual Review of Sociology, 35*, 531–552. https://doi.org/10.1146/annurev-soc-070308-120008.
Chanfreau, J., Lloyd, C., Byron, C., Roberts, C., Craig, R., De Feo, D., & McManus, S. (2008) Predicting Wellbeing. NatCen. Social Research That Works for Society. London: Prepared for the Department of Health. Available at http://www.natcen.ac.uk/media/205352/predictors-of-wellbeing.pdf
Chopel, S. (2012). Culture, Public Policy and Happiness. *Journal of Bhutan Studies, 26*, 82–99.
Conradson, D. (2005). Landscape Care and the Relational Self: Therapeutic Encounters in Rural England. *Health and Place, 11*, 337–348. https://doi.org/10.1016/j.healthplace.2005.02.004.
Crumley, J. (2000). *A High and Lonely Place*. Latheronwheel: Whittles Publishing.
Crumley, J. (2003). *The Mountain of Light*. Latheronwheel: Whittles Publishing.
Crumley, J. (2007). *Brother Nature*. Dunbeath: Whittles Publishing.
Devine-Wright, P., & Clayton, S. (2010). Introduction to the Special Issue: Place, Identity and Environmental Behaviour. *Journal of Environmental Psychology, 30*, 267–270. https://doi.org/10.1016/S0272-4944(10)00078-2.
Diener, E. (1984). Subjective Wellbeing. *Psychological Bulletin, 95*(3), 542–575.
Diener, E., Emons, R. A., Larsen, R. J., & Giffin, S. (1985). The Satisfaction with Life Scale. *Journal of Personality Assessment, 49*, 71–75.
Dluhosch, B., & Horgos, D. (2013). Trading up the Happiness Ladder. *Social Indication Research, 113*, 973–990. https://doi.org/10.1007/511205.
Dolan, P. (2014). *Happiness by Design: Finding Pleasure and Purpose in Everyday Life*. London: Penguin Books.
Dorson, R. M. (1968). *The British Folklorists: A History*. London: Routledge and Kegan Paul Ltd.
Esposito de Vita, G., Bevilacqua, C., & Trillo, C. (2013). Improving Conviviality in Public Spaces: The Case of Naples, Italy. *Journal of Civic Engineering and Architecture, 7*(10), 1209–1219.
Fowler, C., & Helfield, G. (Eds.). (2006). *Representing the Rural: Space, Place and Identity in Films about the Land*. Detroit: Wayne State University Press.
Francis, L. J. (1998). Happiness Is a Thing Called Stable Extraversion: A Further Examination of the Relationship Between the Oxford Happiness Inventory

and Eysenck's Dimensional Model of Personality and Gender. *Personality and Individual Differences, 26*, 5–11.

Freeman, H. (Ed.). (1984). *Mental Health and the Environment*. London: Churchill Livingstone.

Frey, B. S. (2012). Happiness and Public Policies: Fundamental Issue Working Paper. No. 2012–2016. CREMA Switzerland. International Expert Working Group on Wellbeing and Happiness. Appointed by the King of Bhutan for United Nations.

Gilbert, M. (1990). Walking Together: A Paradigmatic Social Phenomenon. *Midwest Studies in Philosophy, XV*, 1–14.

Goldenberg, A., & Saguy, T. (2014). How Group- Based Emotions Are Shaped by Collective Emotions: Evidence for Emotional Transfer and Emotional Burden. *Journal of Personality and Social Psychology, 107*(4), 581–596. American Psychological Association.

Gordon, J., & Grant, G. (Eds.). (1997). *How We Feel*. London: Jessica Kingsley Publishers.

Gov.UK. (2017). Rights of Way and Accessing Land. Available at https://www.gov.uk/right-of-way-open-access-land/overview. Accessed 30 Aug 2017.

Graham, C. (2012). *Happiness Around the World: The Paradox of Happy Peasants and Miserable Millionaires*. Oxford/New York: Oxford University Press.

Groenewegen, P. P., van den Berg, A. E., de vries, S., & Verheij, R. A. (2006). Vitamin G: Effects of Green Space on Health, Well-being and Social Safety. *BMC Public Health, 6*, 149. https://doi.org/10.1186/1471-2458-6-149.

Gruber, J. J. (1986). Physical Activity and Self-Esteem Development in Children: A Meta-Analysis. In G. Stull & H. Ecjert (Eds.), *Effects of Physical Activity in Children* (pp. 330–348). Champaign: Human Kinetics.

Hallerod, B., & Seldon, D. (2013). The Multi-Dimensional Characteristics of Wellbeing: How Different Aspects of Wellbeing Interact and Do Not Interact with Each Other. *Social Indication Research, 113*, 807–825. https://doi.org/10.1007/s11205-012-0115-8.

Hartig, T., Evans, G. W., Jamner, L. D., Davis, D. S., & Gärling, T. (2003). Tracking Restoration in Natural and Urban Field Settings. *Journal of Environmental Psychology, 23*, 109–123. https://doi.org/10.1016/S0272-4944(02)00109-3.

Hartig, T., Kaiser, F. G., & Strumse, E. (2007). Psychological Restoration in Nature as a Source of Motivation for Ecological Behavior. *Environmental Conservation, 34*, 291–299.

Hay, R. (1998a). A Rooted Sense of Place in Cross-Cultural Perspective. *The Canadian Geographer, 42*(3), 245–266.

Hay, R. (1998b). Sense of Place in Developmental Context. *Journal of Environmental Psychology, 18*(1), 5–29. https://doi.org/10.1006/jevp.1997.0060.
Hills, P., & Argyle, M. (2002). The Oxford Happiness Questionnaire: A Compact Scale for the Measurement of Psychological Well-being. *Personality and Individual Differences, 33*(7), 1073–1082. https://doi.org/10.1016/S0191-8869(01)00213-6.
Hinds, J., & Sparks, P. (2007). Engaging with the Natural Environment: The Role of Affective Connection and Identity. *Journal of Environmental Psychology, 28*, 109–120. https://doi.org/10.1016/j.jenvp.2007.11.001. 2008.
Hinds, J., & Sparks, P. (2011). The Affective Quality of Human- Natural Environment Relationships. *Evolutionary Psychology, 9*(3), 451–469.
Hope- Nicholson, M. (1959). *Mountain, Gloom and Mountain Glory: The Development of the Aesthetic of the Infinite*. Ithaca: Reading.
Howell, A. J., Dopko, R. L., Passmore, H. L., & Buro, K. (2011). Nature Connectedness: Associations with Wellbeing and Mindfulness. *Personality and Individual Differences, 51*, 166–171.
Hung, R. (2010). Journeying Between Home and Nature: A Geo-Phenomenological Exploration and Its Insights for Learning. *Environmental Values, 19*(2), 233–251. Stable URL: http://www.jstor.org/stable/30302348
Hunter, J. (1974). The Politics of Highland Land Reform, 1873–1895. *Scottish Historical Review, 5*(1), 45–68. Proquest. Available at http://searchproquest.com.ezproxy.is.ed.ac.uk/docview/1293160353. Accessed 30 Aug 2017.
Hunter, I. R. (2003). What Do People Want from Urban Forestry? The European Experience. *Urban Ecosystems, 5*, 277–284. Kluwer Academic Publishers.
Ingulli, K., & Lindbloom, G. (2013). Connection to Nature and Psychological Resilience. *Ecopsychology, 5*, 52–55. https://doi.org/10.1089/eco.2012.0042.
Jamie, K. (2015). *The Bonniest Companie*. London: Picador.
Jimenez, A. C. (Ed.). (2008). *Culture and Wellbeing: Anthropological Approaches to Freedom and Political Ethics*. London: Pluto Press.
Jordan, M. (2009). Nature and Self – An Ambivalent Attachment? *EcoPsychology, 1*(1), 26–31. https://doi.org/10.1089/eco.2008.0003.
Jugert, P., Greenway, K. H., Barth, M., Buchner, R., Eisentraut, S., & Fritsche, I. (2013). Collective Efficacy Increases Pro-environmental Intentions Through Increasing Self-efficacy. *Journal of Environmental Psychology, 48*, 12–23. https://doi.org/10.1016/j.jenvp.2016.08.003. 2016.
Kahn Jr., P. H. (1997). Developmental Psychology and the Biophilia Hypothesis: Children's Affiliation with Nature. *Developmental Review, 17*, 1–61. https://doi.org/10.1006/drev.1996.0430.

Kahneman, D., & Krueger, A. B. (2006). Developments in the Measurement of Subjective Wellbeing. *Journal of Economic Perspectives, 20*(1), 3–24.
Kals, E., Schumacher, D., & Montada, L. (1999). Emotional Affinity Toward Nature as a Motivational Basis to Protect Nature. *Environment and Behaviour, 31*(2), 178–202.
Kamitsis, I., & Francis, A. J. P. (2013). Spirituality Mediates the Relationship Between Engagement with Nature and Psychological Wellbeing. *Journal of Environmental Psychology, 36*, 136–143. https://doi.org/10.1016/j.jenvp.2013.07.013.
Kaplan, R., & Kaplan, S. (1989). *The Experience of Nature: A Psychological Perspective*. Cambridge: Cambridge University Press.
Kashdan, T. B. (2004). The Assessment of Subjective Well-Being (Issues Raised by the Oxford Happiness Questionnaire). *Personality and Individual Differences, 36*, 1225–1232.
Kellert, S. R. (2002). Experiencing Nature: Affective, Cognitive, and Evaluative Development in Children. In P. H. Kahn Jr. & S. R. Kellert (Eds.), *Children and Nature: Psychological, Sociocultural and Evolutionary Investigations* (pp. 117–151). Cambridge, MA: The MIT Press.
Kyle, G. T., Jun, J., & Absher, J. D. (2014). Repositioning Identity in Conceptualisations of Human-Place Bonding. *Environment and Behaviour, 46*(8), 1018–1043. https://doi.org/10.1177/0013916513488783.
Land Reform (Scotland) Act. (2003). Available at https://www.legislation.gov.uk/asp/2003/2/contents. Accessed 14 June 2018.
Layard, R. (2011). *Happiness: Lessons from a New Science*. London: Penguin Group.
Liptrot, A. (2015). *The Outrun*. Edinburgh: Canongate Books.
Little, J. (2012). Transformational Tourism, Nature and Wellbeing: New Perspectives on Fitness and the Body. *Sociologia Ruralis, 3*, 52. https://doi.org/10.1111/j.1467-9523.2012.00566.x.
Little, J. (2015). Nature, Wellbeing and the Transformational Self. *The Geographical Journal, 181*(2), 121–128. https://doi.org/10.1111/geoj.12083.
Lokhorst, A. M., Hoon, C., Rutte, R. L., & Snoo, G. D. (2014). There Is an I in Nature: The Crucial Role of the Self in Nature. *Land Use Policy, 39*, 121–126. https://doi.org/10.1016/j.landusepol.2014.03.005.
Lopes, M. N., & Cananho, A. S. (2013). Public Greenspace Use and Consequences on Urban Vitality: An Assessment of European Cities. *Social Indication Research, 113*, 751–767. https://doi.org/10.1007/511205-012-0106-9.
Lorimer, H., & Lund, K. (2003). Performing Facts: Finding a Way Over Scotland's Mountains. *The Sociological Review, 51*(sup. s2), 130–144. June 2004.

Louv, R. (2005). *Last Child in the Woods: Saving Our Children from Nature-Deficit Disorder*. Chapel Hill: Algonquin Books of Chapel Hill.
Maas, J., Verheij, R. A., Groenewegen, P. P., de Vries, S., & Spreeuwenberg, P. (2006). Green Space, Urbanity and Health: How Strong Is the Relation? *Journal of Epidemiology and Community Health, 60*, 587–592.
Mabey, R. (1980). *The Common Ground: A Place for Nature in Britain's Future?* (2nd Revised ed.). London: Phoenix Books.
Mabey, R. (2005). *Nature Cure*. London: Vintage/ Penguin Random House Group.
Mabey, R. (2010/2014). *A Brush with Nature: Reflections on the Natural World*. BBC Books. Kindle Edition.
MacCaskil, D. (1999). *Listen to the Trees*. Edinburgh: Luath Press Ltd.
MacDonald, H. (2014). *H Is for Hawk*. London: Random House.
MacFarlane, R. (2003). *Mountains of the Mind: A History of Fascination*. London: Granta Books.
Macfarlane, R. (2007). *The Wild Places*. London: Granta Books.
Macfarlane, R. (2013). *The Old Ways: A Journey on Foot*. London: Penguin.
Macfarlane, R. (2016). *Landmarks*. London: Penguin.
Mackenzie, A. F. D. (2003). Re-imagining the Land, North Sutherland, Scotland. *Journal of Rural Studies, 20*(3), 273–287. https://doi.org/10.1016/j.jrustud.2003.11.001. July 2004.
Mackenzie, A. F. D. (2006). 'S Leinn Fhein am Fearann' (The Land Is Ours): Reclaiming Land, Re-creating Community, North Harris, Outer Hebrides, Scotland. *Environment and Planning D: Society and Space, 24*, 577–598. https://doi.org/10.1068/d398t.
Mackenzie, A. F. D. (2013). *Places of Possibility: Property, Nature and Community Land Ownership*. Chichester: Wiley-Blackwell.
Marcus, G. A. (1995). Ethnography in/of the World System: The Emergence of Multi-Sited Ethnography. *Annual Review of Anthropology, 24*, 95–117.
Mayer, F. S., & Frantz, C. M. (2004). The Connectedness to Nature Scale: A Measure of Individuals' Feeling in Community with Nature. *Journal of Environmental Psychology, 24*, 503–515. https://doi.org/10.1016/j.jenvp.2004.10.001.
Mayer, F. S., Frantz, C. M., Bruehlman-Senecal, E., & Dolliver, K. (2009). Why Is Nature Beneficial: The Role of Connectedness to Nature. *Environment and Behaviour, 41*(5). https://doi.org/10.1177/0013916508319745.
McVean, D. N. (1969). *Ecology and Land Use in Upland Scotland*. Edinburgh: Edinburgh University Press.

Miller, P. (Ed.). (1950). *The Transcendentalists: An Anthology*. Cambridge: Harvard University Press.
Milligan, M. J. (1998). Interactional Past and Potential: The Social Construction of Place Attachment. *Symbolic Interactionism, 21*, 1–33.
Moore, E. O. (1981). A Prison Environments Effect on Health Care Service Demands. *Journal of Environmental Systems, 11*, 17–34.
Morita, E., Fukada, S., Nagano, J., Hamajima, N., Yamamoto, H., Iwai, Y., Nakashima, T., Ohira, H., & Sirakawa, T. (2007). Psychological Effects of Forest Environments on Healthy Adults: Shinrin-Yoku (Forest – Air Bathing, Walking) as Possible Method of Stress Reduction. *Journal of the Royal Institute of Public Health, 121*, 54–63.
Morris, N. J. (2003, July). Health, Well-being and Open Space: Literature Review. *OPENspace*. Edinburgh College of Art (Report).
Nisbet, E. K. (2013, June). Results of the David Suzuki Foundation 30x30 Nature Challenge English Survey. Report. Trent University. Available at http://www.davidsuzuki.org/publications/2013/07/23/30x30%20Nature%20Challenge-Final%20report.pdf. Viewed 27 July 2017.
Nisbet, E. K., & Zelenski, J. M. (2011). Underestimating Nearby Nature: Affective Forecasting Errors Obscure the Happy Path to Sustainability. *Psychological Science, 22*, 1101–1106. https://doi.org/10.1177/0956797611418527.
Nisbet, E. K., Zelenski, J. M., & Murphy, S. A. (2009). The Nature Relatedness Scale: Linking Individuals' Connection with Nature to Environmental Concern and Behaviour. *Environment and Behaviour, 41*, 715–740. https://doi.org/10.1177/0013916508318748.
Nisbet, E. K., Zelenski, J. M., & Murphy, S. A. (2011). Happiness Is in Our Nature: Exploring Nature Relatedness as a Contributor to Subjective Well-Being. *Journal of Happiness Studies, 12*, 303–322. https://doi.org/10.1007/s10902-010-9197-7.
Olwig, K. R. (2002). *Landscape, Nature and the Body Politic: From Britain's Renaissance to America's New World*. Madison: The University of Wisconsin Press.
Packer, B. (1995). The Transcendentalists. In S. Bercovitch (Ed.), *The Cambridge History of American Literature. Vol. 2: Prose Writing 1820–1865*. New York: Cambridge University Press.
Park, B. J., Furuya, K., Kaselani, T., Takayama, N., Kagawa, T., & Miyazaki, Y. (2011). Relationship Between Psychological Responses and Physical Environments in Forest Settings. *Landscape and Urban Planning, 102*, 24–32.
Parsons, R. (1991). The Potential Influence of Environmental Perception on Human Health. *Journal of Environmental Psychology, 11*, 1–23.

Pavot, W., Diener, E., & Fujita, F. (1990). Extraversion and Happiness. *Personality and Individual Differences, 11*(12), 1299–1306.

Payton, S., Lindsey, G., Wilson, J., Ohensmann, J. R., & Man, J. (2008). Valuing the Benefits of the Urban Forest: A Spatial Hedonic Approach. *Journal of Environmental Planning and Management, 51*(6), 717–736. Routledge, Taylor and Francis Group.

Pontin, E., Schwannauer, M., Tai, S., & Kinderman, P. (2013). A UK Validation of a General Measure of Subjective Wellbeing: The Modified BBC Subjective Wellbeing Scale (BBC –SWB). *Health and Quality of Life Outcomes, 11*, 150. https://doi.org/10.1186/1477-7525-11-150.

Proshanky, H. M., Fabian, A. K., & Kaminoff, R. (1983). Place Identity: Physical World Socialisation of the Self. *Journal of Environmental Psychology, 3*, 57–83.

Ramblers: At the Heart of Walking. (2017). Available at www.ramblers.org.uk. Accessed 17 July 2017.

Reid, T. (1764/1997). *An Inquiry into the Human Mind on the Principles of Common Sense*. Edinburgh: Edinburgh University Press

Reith, S. (2008). Through the "Eye of the Skull": Memory and Tradition in a Travelling Landscape. *Cultural Analysis, 7*, 77–106.

Samman, E. (2007). *Psychological and Subjective Wellbeing: A Proposal for Internationally Comparable Indicators*. OPHI Working Paper Series. University of Oxford.

Scotland Outdoors. (2017). Available at http://www.scotoutdoors.com/. Accessed 17 July 2017.

Scottish Natural Heritage. (2017). Enjoying the Outdoors. Available at http://www.snh.gov.uk/enjoying-the-outdoors/. Accessed 17 July 2017.

Seligman, M. (2011). *Flourish: A New Understanding of Happiness and Wellbeing – And How to Achieve Them*. London: Nicholas Brealey Publishing.

Shepherd, N. (1977/2011). *The Living Mountain*. Edinburgh: Canongate Books

Shepherd, N. (2014). *In the Cairngorms*. Cambridge: Galileo Publishers.

Shultz, P. W., Shriver, C., Tabanaco, J. J., & Khazian, A. M. (2004). Implicit Connections with Nature. *Journal of Environmental Psychology, 24*, 31–42. https://doi.org/10.1016/S0272-4944(03)00022-7.

Sparks, P., Hinds, J., Curnock, S., & Pavey, L. (2014). Connectedness and Its Consequences: A Study of Relationships with the Natural Environment. *Journal of Applied Social Psychology, 44*, 166–174. https://doi.org/10.1111/jasp.12206.

Stedman, R. (2002). Toward a Social Psychology of Place: Predicting Behaviour from Place-Based Cognitions, Attitude, and Identity. *Environment and Behaviour, 34*(5), 561–581.

Steptoe, A., & Butler, N. (1996). Sports Participation and Emotional Well-being in Adolescents. *Lancet, 347*, 1789–1792.

Stern, M. J., Powell, R. B., & Ardoin, N. M. (2011). Evaluating a Constructivist and Culturally Responsive Approach to Environmental Education for Diverse Audiences. *Journal of Environmental Education, 42*(2), 109–122.

Stewart, D. (1829). *Elements of the Human Mind in The Works of Dugald Stewart.* Cambridge: Hilliard and Brown (Digitalised by Google Books).

Stigsdotter, U. K., Ekholm, O., Schipperijn, J., Toftager, M., Kamper-Jorgensen, F., & Randrup, T. B. (2010). Health Promoting Outdoor Environments – Associations Between Green Space, and Health, Health-Related Quality of Life and Stress Based on a Danish National Representative Survey. *Scandinavian Journal of Public Health, 38*, 411–417. https://doi.org/10.1177/1403494810367468.

Tam, K. P. (2013). Dispositional Empathy with Nature. *Journal of Environmental Psychology, 35*, 92–104.

The List. (2009). Get Healthy and Get Outdoors and Go for It with the Visit Scotland Adventure Pass…. Available at https://www.list.co.uk/article/15387-get-healthy-get-outdoors-and-go-for-it-with-the-visitscotland-adventure-pass/. Accessed 17 July 2017.

The Network of Wellbeing. (2017). Available at http://www.networkofwellbeing.org/. Accessed 17 July 2017.

The OECD. (2013). *Guidelines for Measuring Subjective Wellbeing.* OECD Publishing. Available at http://dxi.doi.org/10.1787/9789264191655-en

The Scottish Outdoors Access Code. (2005). Public Access to Scotland's Outdoors: Your Rights and Responsibilities. Available at http://www.snh.org.uk/pdfs/publications/access/full%20code.pdf. Accessed 30 Aug 2017.

The Scottish Poetry Library. (2017). Available at Nature tags www.scottishpoetrylibrary.org.uk/poetry/tags/nature. Accessed 17 July 2017.

The Scottish Waterways Trust. (2016). Get Outdoors in 2016. Available at http://scottishwaterwaystrust.org.uk/4572-2/. Accessed 17 July 2017.

Thin, N. (2005). *Happiness and the Sad Topics of Anthropology.* WeD Working Paper 10. ESRC Research Group on Wellbeing in Developing Countries. Available at https://www.researchgate.net/publication/237295096_Happiness_and_the_Sad_Topics_of_Anthropology. Accessed 14th Aug 2017.

Thompson-Coon, J., Boddy, K., Stein, K., Whear, R., Barton, J., & Depledge, M. H. (2011). Does Participating in Physical Activity in Outdoor Natural Environments Have a Greater Effect on Physical and Mental Wellbeing than

Physical Activity Indoors? A Systematic Review. *Environmental Science & Technology, 45*, 1761–1772. https://doi.org/10.1021/es102947t.
Tinkler, L., & Hicks, S. (2011). *Measuring Subjective Wellbeing*. Office for National Statistics, United Kingdom. Available at https://pdfs.semanticscholar.org/faf7/c4986f21b4569d82f1c87045508ae22551fb.pdf. Accessed 27 July 2017.
Trevor-Roper, H. (2014). *The Invention of Scotland: Myth and History*. The Literary Estate of Lord Dacre of Glanton. New Haven: Yale University Press.
Tsunetsugu, Y., Park, B.-J., & Miyazaki, Y. (2010). Trend in Research Related to Shinrin-yoku (Taking in the Forest Atmosphere or Forest Bathing) in Japan. *Environmental Health and Preventative Medicine, 5*, 27–37.
Tuan, Y. F. (1979). *Space and Place*. London: Edward Arnold Publishers Ltd..
Turner, V. (1967). Betwixt and Between: The Liminal Period in Rites de Passage. In *The Forest of Symbols*. New York: Cornell University Press.
Turner, V. (1969). *The Ritual Process – Structure and Anti-structure*. Suffolk: Richard Clay (The Chaucer Press Ltd).
Turner, V. (1970). *The Forest of Symbols: Aspects of Ndembu Ritual*. Ithaca: Cornell University Press.
Turner, V. (1974). *Dramas, Fields, and Metaphors*. Ithaca: Cornell University Press.
Ulrich, R. S. (1979). Visual Landscapes and Psychological Well-being. *Landscape Research, 4*, 17–23. https://doi.org/10.1080/01426397908705892.
Ulrich, R. S. (1981). Natural Versus Urban Scenes: Some Psychophysiological Effects. *Environment and Behaviour, 13*, 523–556. https://doi.org/10.1177/0013916581135001.
Ulrich, R. S. (1983). Aesthetic and Affective Response to Natural Environment. In I. Altman & J. F. Wohlwill (Eds.), *Human Behaviour and Environment: Advances in Theory and Research, Behaviour and the Natural Environment* (Vol. 6, pp. 85–125). New York: Plenum Press.
Ulrich, R. S., & Parsons, R. (1992). Influences of Passive Experiences with Plants on Individual Well-Being and Health. In D. Relf (Ed.), *The Role of Horticulture in Human Well-being and Social Development* (pp. 93–105). Portland: Timber Press.
Ulrich, R. S., Simons, R. F., Losito, B. D., Fiorito, E., Miles, M. A., & Zelson, M. (1991). Stress Recovery During Exposure to Natural and Urban Environments. *Journal of Environmental Psychology, 11*, 201–230. https://doi.org/10.1016/S0272-4944(05)80184-7. 1992.

Van den Berg, A. E., Maas, J., Verheij, R. A., & Groenewegen, P. P. (2010). Green Space as a Buffer Between Stressful Life Events and Health. *Social Science & Medicine, 70*, 1203–1210. https://doi.org/10.1016/j.socscimed.2010.01.002.
Van Herzele, A., & de Vries, S. (2012). Linking Green Space to Health: A Comparative Study of Two Urban Neighbourhoods in Ghent, Belgium. *Population and Environment, 34*(2), 171–193. https://doi.org/10.1007/s11111-011-0153-1.
Vergunst, J. (2013). Scottish Land Reform and the Idea of 'Outdoors'. *Ethnos, 78*(1), 121–146. https://doi.org/10.1080/00141844.2012.688759.
Ward-Thompson, C. (2007). Chapter 3 – Playful Nature: What Makes the Difference Between Some People Going Outside and Others Not? In C. Ward-Thompson & P. Travlou (Eds.), *Open Space People Space* (pp. 23–37). Oxon/New York: Taylor and Francis.
Ward-Thompson, C. (2008). The Childhood Factor: Adult Visits to Green Places and the Significance of Childhood Experience. *Environment and Behaviour, 40*(1), 111–143. https://doi.org/10.1177/0013916507300119.
Ward-Thompson, C. (2010). Landscape Quality and Quality of Life. In C. Ward-Thompson, P. Aspinall, & S. Bell (Eds.), *Innovative Approaches to Researching Landscape and Health*. Abingdon: Routledge.
Ward-Thompson, C. (2011). Linking Landscape and Health: The Recurring Theme. *Landscape and Urban Planning, 99*(2011), 187–195.
Ward-Thompson, C., & Travlou, P. (Eds.). (2007). *Open Space People Space*. Oxon/New York: Taylor and Francis.
Wilson, E. O. (1984). *Biophilia*. Cambridge, MA: Harvard University Press.
Wong, J. L. (2007). Culture, Heritage and Access to Open Spaces. In C. W. Thompson & P. Travlou (Eds.), *Open Space People Space*. Oxon/New York: Taylor and Francis.
Zelenski, J. M., & Nisbet, E. K. (2012). Happiness and Feeling Connected: The Role of Nature Relatedness. *Environment and Behaviour, 46*(1), 3–23. https://doi.org/10.1177/0013916512451901. 2014.
Zevnik, L. (2014). *Critical Perspectives in Happiness Research: The Birth of Modern Happiness*. Springer International Publishing, e-Book. https://link.springer.com/book/10.1007%2F978-3-319-04403-3.
Zhang, J. W., Howell, R. T., & Iyer, R. (2014). Engagement with Natural Beauty Moderates the Positive Relation Between Connectedness with Nature and Psychological Wellbeing. *Journal of Environmental Psychology, 38*, 55–63. https://doi.org/10.1016/j.jenvp.2013.12.013.

2
A Transdisciplinary Ethnography

Jo Vergunst has carried out significant empirical ethnographic research on walking within the Scottish landscape. His focus has been on the body, locomotion and rhythm, and the effects of the body in the landscape on social understanding.[1] He has carried out work in the Cairngorms, Orkney Islands, and Aberdeenshire,[2] considering the perception and cultural representations of the Scottish landscape and the movement of the body through these landscapes as political. Similarly, Vergunst and Arnason[3] considered the moving body within the natural landscape with an interest in mobility and 'place making,'[4] and again Vergunst and Vermehren[5] empirically explored, via arts practices in the landscape, walking, and cycling, the moving body and the implications of shared understanding of place. The themes within these articles were never being made explicit in relation to wellbeing, but nonetheless were implicit with themes such as shared understanding, the body, place making, immersion, and meaning making.[6]

Bell et al.[7] set out to discover how people experienced 'blue-space'[8] in everyday coastal interaction with an interest in wellbeing. Though not entirely ethnographic, instead a mixed methods study, it does allow for qualitative accounts of the local community's interaction with seascapes. They also highlight the fact that much of the previous research within the

R. Crowther, *Wellbeing and Self-Transformation in Natural Landscapes*,
https://doi.org/10.1007/978-3-319-97673-0_2

remits of wellbeing and nature is limited to green space, urban woodlands, parks and gardens, and so on.[9] Bell et al. had an interest in felt bodily sensations. The group's methods included recruitment by screening questionnaires, the participants carrying accelerometers to measure physical activity, and GPS tracking devices in order to measure location. The data from these were used to create 'geo-narrative base maps' detailing where people went, how active they were when there, and how long they spent there. These maps were then used to guide in-depth interviews followed by go-along interviews[10]—again, distracting from the straightforward spoken and observed experience of individuals in nature encounter.

I find Bell et al.'s methods problematic. In terms of being able to capture the lived, in-the-moment experience of these individuals, their methods are missing the point. The pace at which people walk, the areas to which they walk, and how active they are in these areas detracts from the actual experience of being in these spaces. They also negate to comprehend how wellbeing is associated with these experiences. Though the group carried out in-depth 'go-along' interviews, they have gone to little length to limit researcher presence and intervention in encounter. These surplus methods distract from the simple question: how do people experience these spaces and why? What Bell et al.'s empirical research does tell us is that within this group, people had three types of agenda over activity: 'achieving experiences,' 'immersive experiences,' and 'social experiences'—three quite distinct agendas that have resurfaced within my own research. Another of the group's findings that resonate with my own is that, for individuals, 'the presence of people perceived to share similar values and preferences may enhance opportunities for positive therapeutic encounter.'[11] Within my own research, I have found that not only may this be the case but that people will seek to take part in outdoor activities with people that they perceive to be similar to themselves. This aids in the experience being considered as transformational. I will return to this when I discuss self-verification.

Recognising one's personhood, as Cerulo[12] suggests as necessary when defining the self, requires that one's biography and life situation is understood.[13] Understanding someone's experience, self in relation to others, and group dynamic requires the same. This becomes muddied when we consider that for some the social, or interactive at least, is made up of not only the human. What is rather unique within this research project is that it has approached the relationality between the self, group, and non-human in relation to wellbeing. This project has found itself to be more progressive due to the fact that some of my informants may or may not perceive the non-human as an active agent within this social network.

Relationality

> every convincing analysis of outdoor education … comes up with some form of the "self – others – nature" triangle.[14]

The three subjects of self, other, and nature, or non-human or environment are a common grouping for attempted understanding; what Quay[15] refers to as 'the trinity' and a 'way of framing and understanding experience in outdoor education.'[16] However, he too questions whether this can be 'the final word on understanding experience.'[17]

What is not clear in approaching this trinity is how in fact each of the three factors relate in order to contribute to overall experience. What lies between each that so significantly impacts these excursions? How do the three subjects relate to one another? This trinity is suggestive of a network of sorts not unlike Latour's.[18] However, it seems, within the remits of outdoor education, to be considered in a much more pragmatic manner

> through exposure to the outdoor setting individuals learn about their relationship with the natural environment, relationships between the various concepts of natural ecosystems, and personal relationships with others and their inner Self.[19]

There have been few qualitative accounts of group relational encounters with natural space in the United Kingdom and less that were purely ethnographic. Conradson[20] carried out empirical research within a residential elderly centre that emphasised engagement with the natural environment around about it. He wished to approach the relational dynamics of experience, and how therapeutic effects emerge.[21] His work was focused on mobility and care. He too highlighted the emphasis of previous research on therapeutic landscape 'to describe the way in which places become implicated in processes of healing or health enhancement.'[22] He also states that less consideration has been given to these relational dynamics.

Conradson approached his research by engaging with ecological formulations of place and relational notions of selfhood.[23] He was not interested in physiological or biomedical healing; he was however interested in embodied encounters that would later be interpreted and considered alongside the wider context of each individual. Additionally, Conradson acknowledges the potential for ecological perspectives and an attention given to 'the bio-physical environment'[24] to extend the analysis of the self within psychoanalysis and psychotherapy. This potential is found in expanding the normally exclusive psychosocial consideration of relations. Conradson did not however consider the non-human as potentially agentic within these experiences, something that is hinted at within the work of Ingold in his extensive work on networks, the body, walking, and place.[25] For Cerulo, 'sociological analyses of social interaction have been primarily directed toward human-to-human exchange'[26] and 'some are actively questioning such human-only restrictions.'[27]

Amongst the material semioticians, of most significance within the literature with regard to human-nature relationships are perhaps both Actor-Network Theory (ANT)[28] and Non-Representational Theory (NRT).[29] If one wants to begin to understand these relational subjects (the self, group, and non-human), these theories are important to grasp. Regardless of whether one wishes to engage fully, they have guided my own enquiry significantly prior to the field.

> All members – human and nonhuman – can make things happen.[30]

Latour[31] believes that social actors decipher the meaning of all things, rather than meaning being deciphered by overarching roles and structures. By this, he means that social actors determine shared meanings and define objects and the way that we interact with the non-human and how the non-human could be said to interact with us.

For Latour and those who champion ideas of ANT,[32] there are two types of power: power in 'potentia' (the potential to have power) and power in 'actu' (the actual power of things, intrinsically). For Latour,[33] individuals cannot have power nor cause others to act intrinsically. The meaning given to other (non-human and human) is what gives other a sense of power. Energy is given to material things through our collective deciphering of the meaning of things (what he refers to as 'translation'). This is a distributed capacity within the 'network.' Within the network, power is attributed or distributed depending on associated meaning.

Bennet[34] acknowledges the agency of assemblages, something that ANT may not recognise. It could be noted perhaps though, if we heed Latour's discussion of power dynamics and meaning,[35] that within the actor network a non-human may have intention if bestowed with the power of intention by the collective. This may be tied up in language, semiotics, and more relevantly, anthropomorphism. The true belief in intention though is somewhat complex and diverse across groups.

ANT aims to clarify that connections between all things within an ecology (described as 'a network'), whether social, living, or material, have an effect on experience within a given environment. Of course, all ecological analysis is about connections within a given ecology; however, ANT advocates were among the first contemporary social scholars to grant equal status to all actors, within an environment as affective social actors. As the theory's label suggests, every actor *acts* in order that things may occur within this network. ANT considers, as Law[36] does, that all within a given ecology have the ability to cause affect—as a researcher looking at the experience of natural landscapes I agree with this statement; however, acknowledging the effectiveness of the non-human would not allow the non-human any kind of will or intention. A key component of ANT is that it denies the intentionality of the non-human. This book however, due to what was articulated within the field regarding how the non-human is perceived by my informants, explores the non-human as intentional.

Thrift,[37] similarly to Latour and company, considers that accounts of the social (or what may be considered as social) are undefined. According to Thrift, 'much of human life is lived in a non-cognitive world.'[38] For me, this is a concise way to describe such a complex set of ideas as in NRT. Much of our social world is non-representational, that is it cannot be read as a series of signs instead it is 'about mechanics of space brought about by the relation between bodies and things.'[39] If I were to translate Thrift's statement, the reader may well see how NRT may have been helpful to my research prior to the field: elements of the environment—the material (living and non-living) and the body collectively produce a sense of the body within situ. Like in ANT, all 'things' or actors influence, equally, experience.

These concepts were worth coming to grips with ahead of my fieldwork as they encouraged me to approach the material, the elements of environment, and the body holistically, as well as the articulation of experience. I would ethnographically explore not only what was said and seen but I would attempt to understand phenomena at play across and between all actors. This was helpful as these ideas allowed me to consider what it may be within encounter that is intangible. These theories allowed me an openness of exploration when entering the field. It was Thrift's concept of the 'body stance'[40] in which he speaks of a sense of expectation within bodily encounter that I felt *could* be pertinent to an immersion in nature and how it is experienced. However, as an ethnographer I was looking to understand not only the bodily experience but also the cognitive and explicitly representational (the spoken, performative, and behavioural). As an ethnographer, I was looking for signs to comprehend. Thrift's concepts began to hinder how I might engage with fieldwork.

Within human geography, Thrift's work in relation to the body is key; however, as my research set out to holistically represent the experience of my informants, this abstraction became a distraction from the reality of these qualitative experiences. Whilst NRT seemed to focus on the external of experience, I was also interested in the internal and so turned to the cognitive sciences. An ethnographic research into the experience of groups in natural landscapes needed to be representational. It also needed to engage with the minds of my informants since this research explicitly set out to understand the transformational qualities of these excursions: emotional, mental, behavioural, and experiential. It was not to be an exercise in abstraction.

In the past, with theorists such as George Herbert Mead[41] and Erving Goffman,[42] a self or sense of it was limited to humans.[43] This was because 'nonhumans lack[ed] the "I" so critical to the reflexive process.'[44] It was considered that the non-human does not have the capacity to consider themselves as part of any social network. Additionally, the non-human supposedly lacks the linguistic communicability[45] of those who may be deemed intentional. This however, was not the belief of many of my informants. By engaging with ANT and NRT, I could prepare myself for the consideration that all material objects, as well as living beings, within these scenarios may contribute to the experience as a whole. This consideration was useful pre-field as it allowed me to be open to not only relationships between groups but to relationships between humans and material things. ANT and NRT, however, became merely abstraction as my fieldwork continued. This discourse in fact bore no resemblance to what was being explained to me regarding experience by informants. This was for two reasons: either the concepts were in no way relevant to how informants considered their experiences or perhaps more pertinently, in relation to ANT, some informants did in fact consider the non-human as having some kind of power or intentionality within their own experience. This was something that ANT enthusiasts would deny the material actor. Some informants did consider the experience as made up of a kind of network, whilst others viewed the non-human as intentional mediators over their own experience, leading me down a different path within enquiry.

Mental Health

Though this research did not set out explicitly to inform mental health strategies in Scotland, it did unearth reasons as to why exposure to natural environments in Scotland are considered as mentally transformative. I provide an understanding regarding motivations to engage in outdoor activity in Scotland and discuss the links between motivations to engage with natural spaces and social groups and subsequent felt experiences when doing so. In this, the changing self and group dynamic is exposed. Through ethnographic enquiry, I represent groups that do journey to these spaces due to the perceived benefits of natural interaction predominantly in a mental wellbeing capacity.

Whether labelling oneself as in need of mental health improvement or otherwise, there is a clear link between these engagements and positive perceptions regarding oneself and others. This research outlines, for the individuals with whom I researched, why this is so. Due to an evident lack of qualitative, ethnographic, and experiential accounts of nature and human interaction in the United Kingdom, this book provides new ideas regarding these excursions and relationships to multiple fields engaged in looking at nature, green space, and wellbeing.

Contrary to previous research, my own work has focused explicitly on the experiencing self and group in a relational capacity to the non-human within these natural excursions. I have considered wellbeing without furthering the trope, as discussed previously, of instrumentalising wellbeing within eco-philosophy and pro-environmental care. This will contribute to understandings within this area though instead this research has an empathy towards individuals as not directly accountable for environmental challenges. I focused on encounters with groups who aimed for better wellbeing, *happiness*, personal development, therapy, and transcendence. This was activity that was for the benefit of the self, the group, and the community—with any further benefit to the environment as a secondary factor. I wished to explore the connections and the reciprocity that people spoke of. This research provides experiential accounts that come some way to explaining the phenomenological experience of sharing in natural encounter in Scotland. Previous research has failed to allow purely for an understanding of relationships within these environments for mental wellbeing's sake nor have they allowed for an understanding of simply self and group transformation without further agenda, be it economic, environmental, educational, or political.

I believe that previous research across the multiple theoretical disciplines addressing nature and wellbeing have not begun to empirically support ideas that engagements such as these are deeply phenomenological, ineffable, perhaps innately human and often intangible. This research project, however, came from a sincere desire to understand the human relationship with nature specifically in Scotland and to do this ethnographically. I believe that empirical ethnographic research has a strong ability to comprehend, objectively, subjective experiences involving the body, the mind, the integration of groups, group dynamics, and the relationship to these landscapes and encounters within.

From Drama to the Social

> A tentative definition of performance may be: Ritualised behaviours conditioned/ permeated by play.[46]

> It is necessary to develop a form of transdisciplinarity which entails making connections not only across the boundaries of disciplines, but also between scholarly enquiry and the sphere of tacit and experiential knowledges.[47]

I began with performance theory and associated social theory due to my research background in performance analysis and the theatrical arts. The loose and wide definition from Richard Schechner above indicates two main factors in what may be considered as performance: (1) any ritualised behaviour (perhaps structured activity, I thought) and (2) that adopted play (or reverie, leisure, adventure, and exploration, I wondered). The field was ripe with informal social performances and dramatised moments. The act of leaving the urban environment, the rules and structures that go alongside these excursions, the social roles adopted within small group development, and the opportunity for play, reverie, and informal leisure time all suggested that these excursions could be approached with a nod towards performance analysis. Social performance and the performance of self-identity were at the forefront. My background affords me a specific kind of looking and a tacit understanding of performative behaviours. This framework, alongside others, was considered throughout the research process. This was more than the use of disciplinary theory but instead a way of approaching and understanding the dynamics of relationships within fieldwork.

> Performance is an inclusive term. Theatre is only one node on a continuum that reaches from the ritualisations of animals (including humans) through performances in everyday life – greetings, displays of emotion, family scenes, professional roles, and so on – through to play, sports, theatre, dance, ceremonies, rites, and performances of great magnitude.[48]

For Schechner,[49] performance is an inclusive term, a continuum. This continuum ranges from performance in everyday life (with direct reference to Erving Goffman) to play and leisure and indeed theatrical performance.

This framework however is not linear: 'Each node interacts with the others.'[50] By adapting Schechner's fan, as I have done in Fig. 2.1, I show how a drama analogy is useful here. Where for Schechner performance includes rites, ceremonies, shamanism, crisis, everyday life, leisure, play, art making, and ritualisation, I believe that performance can be observed within shared natural encounter. Like Schechner, I leave performance in everyday life in the centre incorporating role play and social performance within the groups that I studied. This is inclusive of the way each informant performed his or her identity. Within the categories of performance within these scenarios, I also add performative liminal moments such as the journey, neo-shamanic practices, mindfulness practice and meditation, small group development dynamics, reverie and idleness, creating with and managing nature, and the creation of structure and boundaries. I believe these things could all be said to be social performance within these excursions. This performance is in the activity itself, the 'framing' or the management of groups and timescales. Considering these excursions as performative is inclusive of all that could be observed. Each category or activity seen below interacts with notions of performance.

> Performances are usually subjunctive, liminal, dangerous, and duplicitous they are often hedged in with conventions and frames: ways of making the places, the participants, and the events somewhat safe. In these relatively safe make-believe precincts, actions can be carried to extremes, even for fun.[51]

Fig. 2.1 An adaptation of Schechner's 'the fan'

I recognise the performative elements of these excursions that separate them from the informant's day to day. Informants deliberately make the journey from urban environments to natural ones. Facilitators are careful in their planning of each excursion and session. They differentiate spaces and times within excursions from those before and after. We could say that they intend to 'frame' these excursions as separate from informants' day-to-day experience within urban environments. We could also say that they make the places, participants, and events safe.

Framing

Turner's ideas regarding social dramas within his 1987 text *The Anthropology of Performance* link his speculations on ritual drama directly to Goffman's belief that 'all social interaction is staged [...].'[52] He discusses Goffman's[53] notion of framing and social performance, commenting specifically on Schechner's discussion of his own ideas. Turner insists that convincing dramatic presentation entails a 'balance between contrivedness and spontaneity.'[54] This could indeed be said of the group excursions that I witnessed. The journeys are planned, often meticulously yet within all case study group excursions there was an inclination towards play, exploration, and spontaneity. I will discuss this again when I speak of structure and anti-structure in Chap. 5. For now, it is important to know that all case study groups 'framed' their excursions. Each excursion had clear structures and timescales and facilitated play, and each excursion in natural landscapes began and ended in urban areas.

Schechner suggests the term 'nest'[55] to refer to the safe space created in performance. The 'nest' is a space and time separated from the everyday much like within these excursions. The case study groups that I have worked with made excursions to these natural spaces for various reasons, but each demanded a space for personal development, personal 'work,' and mental health improvement. Each excursion encompassed ambiguity and uncertainty for informants. As framed events, they were made safe, where informants could feel comfortable taking risks, subverting norms, and adapting to new structures, spaces, and social groups. Excursions displayed all the makings of a social performance for observation and analysis.

> Special rules exist, are formulated, and persist because these activities are something apart from everyday life. A special world is created where people can make the rules, rearrange time, assign value to things, and work for pleasure. This 'special world' is not gratuitous but a vital part of human life. No society, no individual, can do without it. It is special only when compared to the 'ordinary' activities of productive work.[56]

Within performance gatherings, there is a basic structure of 'gathering/ performing/ dispersing'[57]; we see again that this is not unlike how my case study group excursions were structured. They gathered prior to leaving the urban, they participated within the excursion activity within the natural landscape, and then they dispersed, that is they returned to their home life. Schechner's 'nest,' is created by the performance space or 'frame' and through gathering in a specific time and place. It is an agreement between all participating parties to gather, to perform and 'to disperse when the performance is over.'[58] Though informants may not have agreed to perform, they will have agreed implicitly to take part in the social activity—an everyday social performance. They will also disperse in order to return home after this performance.

Ways of Looking

Schechner[59] talks about 'play acts' as being measured against six templates: structure, process, experience, function, ideology, and frame—all key 'templates' through which ethnographers may also observe and describe acts. For Schechner,[60] the first template is 'structure' or the relationship between events. For my purpose, I took this 'relationship between events' and considered the relationships between meetings, excursions, between case studies and further outdoor events within personal narratives. The second template is 'process,' considering how play acts are generated and what the phases of development may be. Here I considered the planning, preparation, gathering of kit, and training in things such as pitching a tent and map reading. The third template is 'experience'—in other words: 'What are the feelings and moods of those participating as players, directors, spectators, observers?'[61] Instead I considered the moods and feelings of participants, facilitators, individuals, and finally myself as an observer. In fact, a direct simile drawn by Schechner: 'while an observer is, for example, a scholar who in studying a phenomenon attempts to maintain a non-involved distance from it.'[62]

Schechner's fourth template is 'function.' He asks, what purpose does the play serve?[63] And I asked what purpose did these activities serve, for whom and why? The next, 'ideology,' considers what values do plays enunciate, propagate or criticise?[64] And whether these values are the same for 'every player, director, spectator and observer.'[65] In a similar manner, I asked what values do groups and organisations hold up, what ethos do they abide by, and what are their agendas and aims? I ask how are these in comparison to the others? How do they relate? Do all informants hold the same values and if not how are they different in relation to these activities. Finally, Schechner's sixth template, 'frame' is intended to aid in considering how 'the players, directors, spectators, and observers know when the play act begins, is taking place, and is over.'[66] I have used this template to decipher that in fact these excursions are framed in much the same way as a performance. The individuals 'know when the play [or excursion experience] begins' because these events are framed, as I have discussed above.

Considering Schechner's[67] 'play acts,' that are measurable against these six templates, enabled me to look at my groups and activities with the understanding that they too were social performances and therefore could be analysed in a similar manner. The templates that Schechner outlines are not unfamiliar to ethnographers.

> In contrast to multi-disciplinarity – in which disciplinary specialists work together maintaining their disciplinary approaches and perspectives – and interdisciplinarity – in which areas or intersections between disciplines are investigated by scholars from two or more areas – *transdisciplinarity has been described as a practice that transgresses and transcends disciplinary boundaries.*[68]

Transdisciplinarity is about moving beyond disciplinary boundaries. In my introduction, I stated that I deliberately avoid overacknowledging the different disciplines from which ideas emerged pre-, during, and post-fieldwork. Within my research, I allowed for the merging and influence of various cognitive, social, and cultural disciplinary understandings. I do not overcompensate for lack of specific discipline.

Russell et al.[69] believe that the need for transdisciplinary research is due to 'the education of the populace and the social distribution of knowledge [...] no longer the exclusive domain of the experts.'[70] The public now have an influence on research. Russel et al. define this change in research strategy as a changing 'knowledge production landscape'[71]—a weakening of the authority previously held by science and academia and a 'movement inside and outside of academia challenging the scientific knowledge and promot[ion of] respect for diverse lay and indigenous knowledges.'[72] They ask: 'Where is the place of people in our knowledge?'[73] Similarly, Horlick-Jones and Sime ask:

> How can we embody the active ways in which people make sense of their worlds [?][74]

Having no allegiance to any one discipline, I am in a strong position in that I am able to consider experience from multiple theoretical perspectives. For Horlick-Jones and Simes,[75] within research there is a necessity to approach our subjects from interpretive perspectives, cultural perspectives, and behavioural perspectives. I have been able to approach my research with the vantage point of several disciplines yet have not held to only one. I considered interpretive perspectives and how informants defined their own experiences within these environments and groups—with ethnographic methods allowing me to delve into subjectivity. I considered cultural perspectives: the social norms and structures surrounding each individual and group. I tried to understand their cultural, social, and economic backgrounds as well as their normal environmental influences and how this may influence their experience of these excursions and their motives for participating. Finally, I observed people's behaviours and how they appeared to be influenced by the environment and others within these scenarios. Some might say that this is in fact standard practice of ethnographic research, yet few ethnographers might refer to themselves as, or engage with their own research as though they are, transdisciplinary.

According to the definitions provided by Russell et al.[76] and Horlick-Jones and Sime,[77] throughout the research process, I became transdisciplinary. In doing so, I attempted to gain a holistic understanding of my research area and what I observed within the field. I 'acknowledge[d] multidimensionality' as called for by Thompson Klein.[78] Transdisciplinarity needs creativity,[79] it 'replaces reduction with a new principle of relativity.'[80] Transdisciplinarity is 'the science and art of discovering bridges between different areas of knowledge and different beings [...] to permit genuine dialogue.'[81] Transdisciplinarity or 'border work' as so called by Horlick-Jones and Sime[82] cuts 'across the boundaries of orthodox disciplinary knowledge.'[83] To be transdisciplinary and ethnographic involves an openness and flexibility to disciplinary theory and previous explanation of social phenomena. I believe that it also requires a development of tacit understanding and a responsive willingness to take part in activities: observant participation.

In a talk entitled 'Anti-disciplinary Interdisciplinarity' in Spring 2015, Tim Ingold suggested that ethnographers must 'know from the inside.' He addressed a sitting-room full of pre-fieldwork anthropology students in Aberdeen and called for methods capable of bending and deforming. He called for methods to take on the characteristics of the world in which we seek to understand: methods capable of becoming responsive and correspondive. Before I entered the field, this notion inspired me. Since, I have read this talk in the form of a book chapter[84] and found that the themes of which he writes were indeed evident to me within my fieldwork. There was something of preparing for the field, being ready to be flexible within the field, and realising that this was how I must *play* it in action within the field that left me open to transdisciplinary modes of thought.

I found, within fieldwork, that often to be involved and capable of taking on roles within case study groups meant calling on different knowledges from my own past, some implicit and some tacit. By this I mean that some of the knowledge on which I drew was formal and some was informal (social) knowledge. This brought further perspectives into the mix. I was required to lead workshops and to provide additional support to people who required it in terms of behavioural needs or learning difficulties. I drew on creative work with vulnerable people in my past to engage informants. I drew on my creative side so that I could join participants in learning and making. I remembered rock climbing with my father as a child and walking long distances. I remembered map reading and orienteering and pitching a tent or starting a fire. Not only did remembering (and engaging with these knowledges) and learning new things with informants increase my understanding through participation, it also allowed a reflexivity I may not have achieved otherwise. Full participation also allowed a space to develop relationships with informants that may not transpire if one was merely observing. For Ingold,[85] 'to practice participant observation [...] is to join in correspondence with those amongst who we study.'[86] I would agree with the sentiment and insist that it is observant participation that allows for this correspondence.

Ingold challenges the 'inter' in interdisciplinarity. 'Inter' etymologically has connotations of between-ness, meaning that within interdisciplinarity, enquiry is still closed within respective disciplinary territories. This research was not interdisciplinary, it was not bound by disciplinary territories. Additionally, Ingold[87] stated that interdisciplinarity is valued 'precisely in the opportunities it affords to think holistically, in terms of the totality of joined up knowledge.' He believes that this impedes research scholarship. This is because research becomes 'self-conscious,' and it insists on the communication of otherwise 'closed disciplinary identities.'[88] It was necessary for me to adopt a transdisciplinary research strategy towards my research to comprehend what I was finding in the field with groups. I sought to be '*anti-disciplinary*.'

I wanted to 'undo rather than reinforce these [disciplinary] boundaries.'[89] I wanted to 'converge the reality and the meditative'[90] and to represent both as important to this research and the understanding of experience, the imagined, remembered, or anticipated within these ephemeral and intangible experiences of natural landscapes. Ingold had spoken to us in 2015 of the detriment of diminishing experiences within boundaries and frameworks as this 'ceases wonder and astonishment.' I intended to allow the flexibility to explore and represent all facets and to do this in whatever way presented itself within the field. I needed to be responsive and to maintain a commitment to truth, regardless of whether that *truth* was merely perceived by each informant. I set out to approach both the physical and literal and metaphorical and psychological experience of group excursions into natural space. When with these groups, my dominant method was to simply *be* with the group, to do as they do, and integrate as a researcher. It was important for me to be without discipline, to be within the field without disciplinary normative structures and theoretical frameworks guiding my behaviour or line of enquiry. I viewed these excursions as framed and performative and intended to remain flexible in my participation. I allowed emergent themes to guide my theoretical exploration and analysis. This research was transdisciplinary as I approached my research subjects and environments intending to understand from between disciplines, beyond discipline in tacit relationships and understanding, and by remaining responsive to interpretive, cultural, and behavioural perspectives.

In order to comprehend the data, found within the field, it became essential to develop a secondary research strategy to engage with the multiple disciplines that would allow for understanding these complex and intangible experiences. The latter were not only physical but physiological and psychological and often, reaching the informants' true experience as an ethnographer was problematic. It became necessary for me to become adept at reading and applying theory from across multiple disciplines to comprehend my data. I therefore developed a transdisciplinary secondary research strategy.

The above diagram (Fig. 2.2) shows the phenomenological relationships of interest. Each area of interest sits somewhere between 'nature' and 'culture.' It shows how the intrapersonal (the self) interrelates with the interpersonal (the group) and the ecological (the natural landscape and other environmental factors). The reader will see that between each of these actors there is a boundary, that I have labelled 'affect.' Each actor when interrelating in experience affects the other. I argue that within this, there are metaphorical liminal thresholds, discussed later in Chap. 5. These liminal thresholds are represented by the two-point arrows that lie across 'affect.' Each of these thresholds crossed imply transformation, as seen in the centre of the diagram: all three actors affect this transformation. Each threshold has significant implications for the sense of self, the group dynamic, and the perception of the landscape within experience. The self affects the group dynamic. The group dynamic affects the self.

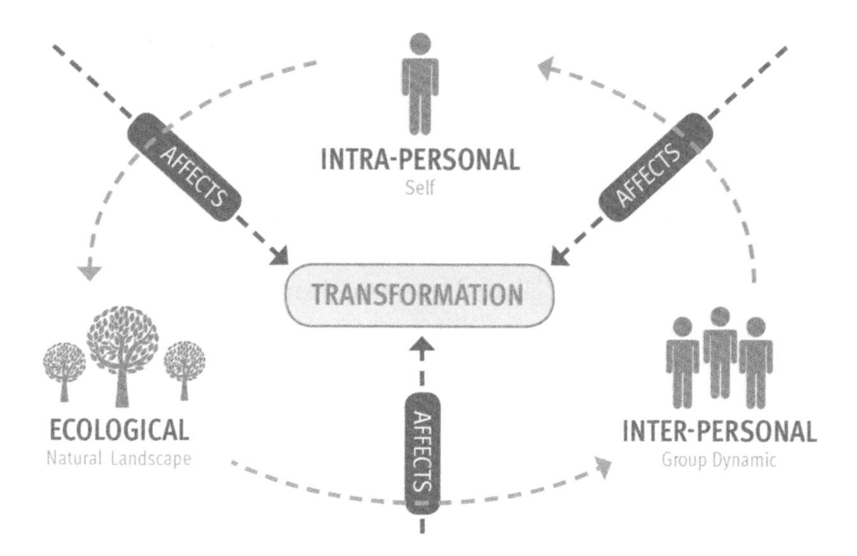

Fig. 2.2 Relationships of interest

The group dynamic affects the perception of and interaction with the natural landscape, and the natural landscape affects the group dynamic. The natural landscape also affects the self as the changing self affects individuals' perception of the natural landscape. Through continuous crossing of metaphorical liminal thresholds in the mind, the group dynamic, and perception of the landscape, transformation occurs.

The scale of transformation varies across individuals and groups. Sometimes this is major intrapersonal and group development both articulated and observed and sometimes subtler intrapersonal changes in the way the group, activities, and landscape are spoken about. Sometimes these transformations are observed through physical behaviours within these circumstances.

The 'interpersonal' subject, or the group as a research subject, required a drawing upon cultural disciplines that focus on the impact of culture upon behaviours and perspectives for it to be understood in terms of its dynamics, its formations, and its contexts. This included looking at constructed ideas and cultural and social contexts of people that may affect these experiences, language and history, and so on. This is to do with how the group is immersed within the land and how they share the space with others, and what happens when they do. Here is where my understanding of abstraction and context can be found. The 'intrapersonal' subject, that is the self, required an understanding of psychological and cognitive disciplines. Finally, the 'ecological' subject required me to engage with ecological, philosophical, and metaphysical ways of understanding the relationship between human and non-human entities. When I moved from the animate to inanimate in analysis, it was arguably metaphysical assumption and highly subjective. Here though I focus on what is explicitly said by informants in relation to their perception of the landscape and to this relationship with the non-human.

Narratives and Abstraction

As an ethnographer, one's 'tools' traditionally only allow for exploration of the spoken and observed amongst informants rather than the intrapersonal, cognitive, or physiological. Luckily within fieldwork, some can articulate and choose to attempt to articulate their experiences. Unfortunately, some either cannot or choose not to articulate these experiences.

Figure 2.3 below addresses this difficulty and references Galen Strawson's 'Diachronic' and 'Episodic' personality types.[91] The metaphorical barriers that prevent the ethnographer from accessing the 'episodics' experience appear between the physical or vocal behaviours, the action seen and the way it is articulated, and the physiological and mental aspects of experience. This is due to the 'episodics' lack of 'narrative self.'[92] When Strawson discusses the 'narrative self,' he explains how some people create conscious and explicit narratives pertaining to themselves and to their experiences: these are 'diachronics.' He also describes other people, 'episodics,' who do not feel the need or are not interested in talking about themselves in this way. He therefore speaks of those who manifest a diachronic self, someone that is helpful to an ethnographer and an episodic self, someone who makes an ethnographers' job just a little bit harder.

The diachronic self has a strong sense of their own narrative or an 'I that is a mental presence now, was there in the past, and will be there in

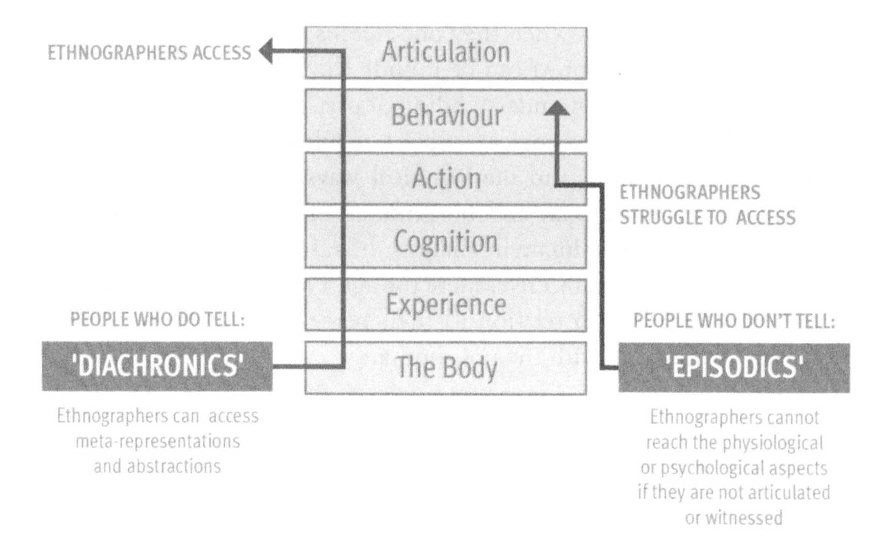

Fig. 2.3 Articulation and experience. ('Diachronics' and 'Episodics,' Strawson 2005)

the future.'[93] Diachronics also create a metarepresentational diachronic self, allowing them to talk about their feelings and mental experiences as well as associating this to their own autobiography. In other words, imagination.[94] This I believe is a similar concept to abstraction, individually differentiated across informants due to cultural context. Culture allows for abstraction beyond the physical here and now, the body in place. Without being able to articulate—to contextualise or abstract ideas in reference to everyday life—thought processes or personal narratives are, to an ethnographer, inaccessible. An ethnographer cannot then access truthful experience and will struggle to defend cognitive assumptions made about an informant's experience.

> Many cultural and social anthropologists not only omit in their studies to take into account the workings of the mind, they are actively hostile to any attempt to do so.[95]

An important thing to think about when speaking to people about their experience is abstraction. This is something I encountered, when struggling with informants who did not say much at all or who could not articulate themselves. This concept became key within my research, in relation to both methods and to the informant's experience. It may link clearly to one's personal narrative and perception of experience. This also, I believe, highlights a useful insight for ethnography as well as an insight into how we may possibly begin to understand individual perceptions of landscapes.

As I have shown when talking about 'diachronic' and 'episodic' personality types, some people do and some people do not abstract ideas. Abstraction allows an insight into an informant's personal narrative. Some people may associate experiences with other experiences, and some people will experience these interactions differently due to their ability to abstract ideas. I found that often these abstract ideas were influential in the way people thought about their experiences. The contexts from which they came and the knowledge that they carried due to past experiences influenced their experience and how they approached their experience. Often this knowledge was as simple as 'a walk is good for me'—they *know* this because they have experienced positive feelings in similar situations before and they are able to then place that understanding within each new interaction. For others, this is not the case.

The discussion of abstraction and of who has the ability to abstract ideas has in the past been controversial, with Bourdieu's[96] portrayal of the subordinate and dominant (elite) classes. He believed that whilst the subordinate classes were merely practical and sensual, the dominant classes were able to abstract ideas, 'revolving around more aesthetic frames of meaning and reference and the concerns of conscious symbolism.'[97] Basil Bernstein, a sociologist specialising in education, agreed with Bourdieu that 'notions of abstraction and concreteness revolve around the degree to which one stands back from one's context.'[98] Interestingly though, within my own case study groups, it seemed to be that those with the ability to abstract and contextualise were often those who spoke themselves about being more immersed, sensually, in the activity. Bourdieu and Bernstein, though, were referring to the hands-on working classes of their age and when they speak of standing back, I imagine they were referring to the dominant classes not involving themselves in the physicality and practicalities of working life. They were not, I suspect, anticipating the rise of activities in nature for personal and social transformation nor the contemporary agendas for immersing oneself in these environments.

Abstraction is often contested and considered as bias, but it is realistic: after all, what are people trying to gain by taking part in these excursions? They have come with a predetermined idea, an abstraction, which will influence choice, affordance, and experience. This may also be influenced by the group facilitators and organisation throughout experience. It became clear that within specific groups, due to this, they developed a collective intention to have specific kinds of experiences. Within other groups, where people did not show (at least not to me as researcher) the ability to abstract, this was not the case. Walks were not a positive experience because not only had individuals had no knowledge (no previous contexts), they were not as able to abstract ideas when describing their experiences. This was not only difficult for me as an ethnographer but also led me to question whether, without context, these interactions had the same outcomes for individuals. In these circumstances, it was the group facilitators who brought with them agendas or aims; whether the intended emotions were felt by participants was another matter. Almost all of the informants from within one case study, for example (The Young Adult's Development Project introduced in Chap. 3), at the end of the programme told me that they still did not know why they were there or what it had *done* for them.

Could abstraction be what it is that distinguishes the understanding and experience of groups. Perhaps it could be the difference between those with prior knowledge and those without?

Those without the ability to abstract the scenario, activity, and landscape often said things like 'I don't like it,' 'I do like being outdoors,' or 'I hate walking.' Those with this ability to abstract however, often compared the activity to something else within their own narratives, knowledge, and experience. They abstracted the thing of which they spoke to incorporate wider themes than those particular to the then and there—that is, they not only considered it in relation to the physical and social now but to something out with that context. Some informants brought emotions and understandings of the situation from elsewhere, and this affected their perception of experience. This contextualising affected their current perception of experience and further understandings. Without this prior context, the in-the-now was only experienced as it was. If a person had no positive experiences to compare, then this experience was not like those with personal or positive context for the excursion. If an informant did not 'know' that a walk might make them feel good, then a walk may not have made them feel good. Rather, these experiences, if only cultural (and not innate) may not have any articulated transformative effect at all unless a context was at first developed. This is further developed within Chap. 4.

I do not believe that those without the desire, or ability, to articulate abstract ideas do not manifest abstract ideas relating to the experience within their minds. This brings us back to the diachronic and episodic self: the episodic self simply does not allow the ethnographer access to these thoughts, whether conscious or not, regarding experience. This suggests that different experiences may not necessarily depend on individual context but instead that they are or are not articulated in particular ways.

A diachronic person, as someone who uses language to abstract their experience, even without prior experience is able to articulate what they are feeling in the moment. However, they will not have a context in which to place this other than the now. If positive feelings within nature are innately human, they may feel these. If they are cultural and based on life contexts, then perhaps they will not. Regardless this new experience will inform future experiences and perceived opportunities within the landscape. This may be true of multiple further experiences. An episodic without prior experience is there in the moment and detached from either previous experience of this kind of scenario or the ability to abstract this current experience. An episodic may abstract ideas within their mind, and this may inform future experience. However, I would have no access to this as an ethnographer. Again, if these experiences are innate, then they may feel similarly to others; however, if they are purely cultural and informed by context they may not.

Both the diachronic and episodic, without previous experience or context may not have the same positive experience as those who associate these activities with their own wider narratives. They may however be able to build this context and compare future excursions to the present. This may be positive, or it may be negative. Those with context often came to excursions with predetermined "knowledge" of the positive benefits of this engagement. Group facilitators and those groups that had this positive context, collectively had an intention to feel positively transformed and often obliged in this endeavour. Those without did not. I suggest that without the context of experience of the outdoors within their own personal narrative, informants experienced these excursions very differently, specifically in relation to their perception of themselves within the group and landscape and how they were affected.

Ethnographic methods allow for an insight into how individuals associate the landscape with their own personal narratives and their personal and social contexts. This is done through conversation, sharing stories and experiences—often brought about through tacit engagement and participation with informants. Through engaging in activities with informants, I was able to witness the abstraction of ideas—the association of activities and the landscape to informant's contexts and thus how they understood these experiences. Where individuals did not articulate ideas in the same manner or for those who were not able (or willing) to abstract thoughts, I was able to question in the moment and to observe, through participation, their behaviours and material and social interactions.

Notes

1. Takano's (2005) work with Inuit communities looking at connection with the land, bonding, and place may also be interesting here. Few ethnographies however aim to have a specific interest in transformative encounter or wellbeing. Takano's work is based in the realm of environmental education and skill sharing.
2. Vergunst (2013).
3. Vergunst and Arnason (2012).
4. See also Sarah Pink (2008).
5. Vergunst and Vermehren (2012).
6. Additionally, on walking and wellbeing, see Morris (2003), Edensor (2000), Duerden (1978), and Wallace (1993). On the reflexive body and different types of walking, see Kay and Moxham (1996). Within sociology see Lewis (2000), discussing the climbing body. Also see Morris (2009) on 'the camper.'
7. Bell et al. (2015).

8. Blue-space refers to coastal areas and seascapes of which my research did engage at points. Within the remit of coastal areas and wellbeing, according to Bell et al., proximity has been associated with general and mental health and increased physical activity (Bauman et al. 1999; Witten et al. 2008 in Bell et al. 2015).
9. Bell et al. (ibid: 56).
10. Bell et al. (Ibid: 58).
11. Bell et al. (ibid: 64).
12. Cerulo (2009).
13. Cerulo (ibid).
14. Walker (1998) in Quay (2013: 143 original emphasis).
15. Quay (2013: 142).
16. Quay (2013).
17. Quay (2013: 142).
18. Latour (2005).
19. Priest (1986) in Quay (2013: 143).
20. Conradson (2005).
21. Conradson (2005: 337).
22. Conradson (2005: 337).
23. Conradson (2005).
24. Conradson (2005: 340).
25. (for e.g., 2004, 2010).
26. Cerulo (2009: 531).
27. Cerulo (2009: 532).
28. Callon et al. (1986), Latour (1987, 1996, 2005) and Law (1992).
29. Thrift (2008).
30. Law 1992, in Cerulo (2009: 535).
31. Latour (1986).
32. Callon et al. (1986), Latour (1986, 1987, 1996, 2005) and Law (1992).
33. Latour (1986).
34. Bennet (2010).
35. Latour (1986).
36. Law (1992).
37. Thrift (2008).
38. Thrift (2008: 58).
39. Thrift (2008: 39).
40. Thrift (2008).
41. Mead (1934).
42. Goffman (1956).

43. Cerulo (2009: 532).
44. Goffman (1956).
45. Goffman (1956).
46. Schechner (1988: 99).
47. Horlick-Jones and Sime (2004: 445).
48. Schechner (1988: xvii).
49. Schechner (1988).
50. Schechner (1988: xvii).
51. Schechner (1988: xix).
52. Schechner in Turner (1987: 74).
53. Goffman (1956).
54. Turner (1987: 149).
55. Schechner (1988).
56. Schechner (1988: 13).
57. Schechner (1988: 189).
58. Schechner (1988: 139).
59. Schechner (1995).
60. Schechner (1995).
61. Schechner (1995: 25).
62. Schechner (1995).
63. Schechner (1995).
64. Schechner (1995).
65. Schechner (1995).
66. Schechner (1995: 25).
67. Schechner (1995).
68. Russell et al. (2007: 460–461, emphasis my own).
69. Russell et al. (2007).
70. Russell et al. (ibid: 464).
71. Russell et al. (2007: 464).
72. Russell et al. (2007: 465).
73. Scholz in Russell et al. (2007: 464).
74. Horlick-Jones and Sime (2004: 442).
75. Horlick-Jones and Sime (2004).
76. Russell et al. (2007).
77. Horlick-Jones and Sime (2004).
78. Thompson Klein (2004: 516).
79. Thompson Klein (2004).
80. Thompson Klein (2004: 516).
81. Thompson Klein (2004: 516).

82. Horlick-Jones and Sime (2004).
83. Horlick-Jones and Sime (ibid: 441).
84. Ingold (2018).
85. Ingold (2018).
86. Ingold (2018: 63).
87. Ingold (2015).
88. Ingold (2015).
89. Ingold (2015).
90. Ingold (2015).
91. Strawson (2005).
92. Strawson (2005).
93. Bloch (2012: 131).
94. Bloch (2012).
95. Bloch (2012: 03).
96. Bourdieu (1984).
97. In Carrier (2001: 246).
98. In Carrier (2001).

References

Bell, S. L., Phoenix, C., Lovell, R., & Wheeler, B. W. (2015). Seeking Everyday Wellbeing: The Coast as Therapeutic Landscape. *Social Science and Medicine, 142*, 56–67.

Bennet, J. (2010). *Vibrant Matter: A Political Ecology of Things*. Durham/London: Duke University Press.

Bloch, M. (2012). *Anthropology and the Cognitive Challenge*. Cambridge: Cambridge University Press.

Bourdieu. (1984). *Distinction*. Cambridge, MA: Harvard University Press.

Callon, M., Law, J., & Rip, A. (1986). *Mapping the Dynamics of Science and Technology: Sociology of Science in the Real World*. Basingstoke: Macmillan.

Carrier, J. (2001). Social Aspects of Abstraction. *Social Anthropology, 9*(3), 243–256.

Cerulo, K. A. (2009). Nonhumans in Social Interaction. *The Annual Review of Sociology, 35*, 531–552. https://doi.org/10.1146/annurev-soc-070308-120008.

Conradson, D. (2005). Landscape Care and the Relational Self: Therapeutic Encounters in Rural England. *Health and Place, 11*, 337–348. https://doi.org/10.1016/j.healthplace.2005.02.004.

Duerden, F. (1978). *Rambling Complete*. London: Kay and Ward.

Edensor, T. (2000). Walking in the British Countryside: Reflexivity, Embodied Practices and Ways to Escape. *Body and Society, 6*(3–4), 81–106. https://doi.org/10.1177/1357034X00006003005.

Goffman, I. (1956). *The Presentation of the Self in Everyday Life*. New York: Random House.

Horlick-Jones, T., & Sime, J. (2004). Living on the Border: Knowledge, Risk and Transdisciplinarity. *Futures, 36*, 441–456. https://doi.org/10.1016/j.futures.2003.10.006.

Ingold, T. (2004). Culture on the Ground: The World Perceived Through the Feet. *Journal of Material Culture, 9*(3), 315–340. https://doi.org/10.1177/1359183504046896.

Ingold, T. (2010). Footprints Through the Weather-World: Walking, Breathing, Knowing. *Journal of the Royal Anthropological Institute* (N.S.), *16*, s121–s139.

Ingold, T. (2015, May). *Anti-disciplinary Interdisciplinarity.* Seminar given at Scottish Training in Anthropological Research. The Burn, Aberdeenshire.

Ingold, T. (2018). *Anthropology and/ as Education*. Oxon: Routledge.

Kay, G., & Moxham, N. (1996). Paths for Whom? Countryside Access for Recreational Walking. *Leisure Studies, 15*(3), 171–183.

Latour. (1986). The Power of Associations. In J. Law (Ed.), *Power, Action and Belief: A New Sociological Knowledge?* (pp. 264–280). London: Routledge and Kegan Paul.

Latour, B. (1987). *Science in Action: How to Follow Scientists and Engineers Through Society*. Cambridge, MA: Harvard University Press.

Latour, B. (1996). On Actor-Network Theory: A Few Clarifications Plus More than a Few Complications. *Soziale Welt, 47*, 369–381. Stable URL: http://www.jstor.org/stable/40878163. Accessed 27 July 2017

Latour, B. (2005). *Reassembling the Social: An Introduction to Actor-Network-Theory*. Oxford: Oxford University Press.

Law, J. (1992). Notes on the Theory of the Actor Network: Ordering, Strategy, and Heterogeneity. *Systems Practice, 5*(4), 379–393.

Lewis, N. (2000). The Climbing Body, Nature and the Experience of Modernity. *Body and Society, 6*(3–4), 58–80.

Mead, G. H. (1934). *The Social Self in Mind, Self and Society*. Chicago: University of Chicago Press.

Morris, N. J. (2003, July). Health, Well-Being and Open Space: Literature Review. *OPENspace*. Edinburgh College of Art (Report).

Morris, N. J. (2009). Naked in Nature: Naturism, Nature and the Senses in Early 20th Century Britain. *Cultural Geographies, 16*, 283–308. https://doi.org/10.1177/1474474009105049.

Pink, S. (2008). An Urban Tour: The Sensory Sociality of Ethnographic Place-Making. *Ethnography, 9*(2), 175–196. https://doi.org/10.1177/1466138108089467.

Quay, J. (2013). More than Relations Between Self, Others and Nature: Outdoor Education and Aesthetic Experience. *Journal of Adventure Education and Outdoor Learning, 13*(2), 142–157. https://doi.org/10.1080/14729679.2012.746846.

Russell, A. W., Wickson, F., & Carew, A. L. (2007). Transdisciplinarity: Context, Contradiction and Capacity. *Futures, 40*, 460–472. https://doi.org/10.1016/j.futures.2007.10.005.

Schechner, R. (1988). *Performance Theory*. London/New York: Routledge.

Schechner, R. (1995). *The Future of Ritual: Writings on Culture and Performance*. London/New York: Routledge.

Strawson, G. (2005). Against Narrativity. In G. Strawson (Ed.), *The Self?* Oxford: Blackwell Publishing.

Takano, T. (2005). Connections with the Land: Land Skills Courses in Igloolik, Nunavut. *Ethnography, 6*(4), 463–486. https://doi.org/10.1177/1466138105062472.

Thompson Klein, J. (2004). Prospects for Transdisciplinarity. *Futures, 36*, 515–526. https://doi.org/10.1016/j.futures.2003.10.007.

Thrift, N. (2008). *Non-representational Theory: Space, Politics, Affect*. London/New York: Routledge.

Turner, V. (1987). *The Anthropology of Performance*. London: PAJ Publications.

Vergunst, J. (2013). Scottish Land Reform and the Idea of 'Outdoors'. *Ethnos, 78*(1), 121–146. https://doi.org/10.1080/00141844.2012.688759.

Vergunst, J., & Arnason, A. (2012). Introduction: Routing Landscape: Ethnographic Studies of Movement and Journeying. *Landscape Research, 37*(2), 147–154. https://doi.org/10.1080/01426397.2012.669955.

Vergunst, J., & Vermehren, A. (2012). The Art of Slow Sociality: Movement, Aesthetics and Shared Understanding. *Cambridge Anthropology, 30*(2), 127–142. https://doi.org/10.3167/ca.2012.300111.

Wallace, A. (1993). *Walking, Literature and English Culture*. Oxford: Clarendon Press.

3

Getting Out, Goethe, and Serendipitous Ethnography

Initially I aim to offer as best I can for the sake of transparency, a concise account of what was done within the field. However, I will do this concisely before moving on to providing a more extensive insight into what this fieldwork realistically entailed: complex metaphysical, emotional, psychological, bodily, meditative, physical, intangible, and social and personal relationships and actions within these natural environments. Methods were undefined and instead came down to *being* within the field, with informants, and within spaces.

Several pilot excursions were embarked upon; most of these were autoethnographic and were carried out pre-fieldwork. They aided in understanding my own subjective disposition within natural landscapes and in interaction with groups. They, more importantly, informed my rationale towards methods: there would be minimal structure to methods, minimal technology, the use of handwritten fieldnotes, and unstructured conversations.

My first pilot study with others was tentative. I was invited, alongside fellow researchers, to a training event in Aberdeen, to an estate surrounded by natural landscape—I capitalised on having a group here to navigate some of my initially thought-out methods. We walked together in the woodlands and by the river. Below are a few of the things that were

R. Crowther, *Wellbeing and Self-Transformation in Natural Landscapes*,
https://doi.org/10.1007/978-3-319-97673-0_3

shared with me on this walk. I have chosen these statements as I believe they highlighted well some of the issues that I may have come to face methodologically later in my research process. I have italicised some intriguing comments:

> I find something a bit *sacred* about it, I'm not talking about *metaphysical*, like, deeply profound…it's breathing. You know? Life, there's so much life to it, *it's life giving*. (Sam)

> The idea of, how does one thing connect to another thing? To me, I call it *spiritual*. Like, I don't know, *unexplainable*. That's just a language I use […] pinning it down is probably something problematic. (Tom)

> I'm taking you here because I felt *something*. *As a child*, my parents had a farm. Now, *I don't like nature* but I almost cried. I ran to it. This is *my castle*. (Robyn)

> It's a huge part of *my story* […] I too was massively depressed with suicidal thoughts […] This is really personal so don't laugh. I find a great peace, like, a huge, peace, like, and I would call it *my soul* is at rest and peace when I see the beauty of nature. Like, it's, it's just an awe of… it makes me feel, insignificant, but not in a negative way. Like… I'm not the centre of *the universe*. (Paul)

From these informal conversations, I heard clearly that these kinds of spaces brought complex ideas to the fore. These were concepts that may not easily be grasped: spirituality, metaphysical aspects, anthropomorphism, connections, ineffability, nostalgia, memory, imagery, and subjective personal relations and associations. When this group left the urban environment and entered this natural landscape, they began to share stories, experiences, and memories. People seemed to open up and converse freely. They spoke of childhood, connection, mental health, associations, and people in their lives. My methods needed to respect this seeming momentary revelation, spontaneity, and openness. These were not themes that I would have easily asked questions about. How would I have known to? They were themes that simply emerged from being there. Through short pilot excursions with groups, and alone, methodological and practical questions emerged as to how I would in fact conduct this research. I was aware that these experiences may have been deeply personal and would demand an empathy and respect towards what may emerge.

Whilst at another group pilot excursion at a small woodland between Edinburgh and Gifford, I was asked to tell 'the story'[1] of my two hours of 'alone time' in the woodlands. I felt incredibly vulnerable and unwilling to share, certainly not open to telling what in reality had surfaced in my thoughts. What arose was deeply personal and would require a little more time in order for me to be open to sharing with the group. This encounter affected all subsequent activities on this excursion due to my fear of having to share again. I was not alone in feeling this distress. I have discovered through conversation with other participants and previous participants of workshops adopting these methods that this kind of sharing, for introverts, may have negative consequences on accurate accounts of experience. This was an experience that has led me to be aware of asking informants to share their experiences immediately post-event.

I had hoped to capture what happens within these transient and ephemeral moments. It would seem, however, as an ethnographer I had a problem.

On another pilot excursion, whilst making my way from Fort William to Inverness by foot through the hills, forests, and lochside, I was struck by several questions with regards to traditional ethnographic methods. The issues that arose were to do with the effect that my documenting throughout experience would have on informants and what effect this would subsequently have upon accurate or representative accounts of these experiences.

Whilst battling with my iPhone in an attempt to document the landscape, I was stuck, again. This digital device was detrimental to my *being within* the situation: alert, reflexive, and critical. Although I had an overwhelming desire to photograph, my digital storage space was full. Distressed, I obsessed over fixing this. I certainly was not within the moment in the landscape or indeed tuned in to the people that I was with. On pondering this, another issue arose. I considered, due to not wanting to disrupt these experiences—the 'flow,'[2] perhaps interviewing informants afterwards may be the answer? I discovered another issue here. This time it concerned memory and the feasibility of accurate retelling or recounting of what was experienced in the moment. This led me to question validity—did it matter if what was recollected was accurate? Or would it merely matter that the informant believed this recollection to be true of the experience had? If I was concerned with the natural space and its effect on groups of people, surely *when* they subsequently recounted that experience shouldn't have mattered? I felt, however, that the act of retelling, recounting, and remembering may have altered this experience each time. This was a concept that reappeared within the field:

> I've noticed recently that sometimes this is actually the space where I'm able to have the headspace [...] while I'm making a spoon or whatever, certain problems that I'm having or certain things that are coming up are actually being processed in this safe space because I'm not alone but I'm alone enough, you know? To safely ask myself some questions or go through some things in my mind and make decisions about what should I do about that thing or how do I really feel about, you know? Where I'm able to come up with clearer answers for myself whereas if I'm alone, alone and I don't know that someone's gonna bring me out of it, or you know? Or miss me if I'm not there, then I will tend to go into escapist mode and I will not address any of my things, because it just feels too overwhelming so occasionally I get these nice moments where I'm here and I start to feel better about whatever is going on in my life and I gain perspective and I think hey, OK! So not everything's perfect but I'm here right now, with all of these awesome people and I'm doing this thing and in an hour I'm gonna get lunch and its gonna taste great... (Holly)

This point made by Holly is important in terms of method. Holly was able to articulate these feelings towards being at the site whilst immersed in the experience. In this though she was also able to reflect. Having engaged with Kahneman and Riis,[3] where they discuss the dichotomy of the experiencing and the remembering self, several decisions with regard to timely points of reflection and conversation were made. As Kahneman and Riis state: 'Retrospective evaluations are [...] less authoritative than reports of current feelings.'[4] In that vein, according to Kahneman and Riis,[5] if a participant-informant is asked about an experience, after encounter, it will be the evaluating self who answers. Whilst on the one hand, I wished to avoid impacting the 'flow' of experience, methods needed to do their best to allow for in-the-moment capturing of the experience. If a decision was made to only acknowledge what is said out with the experience or upon reflection, the memory perhaps may become more tangible, however 'the remembering self is sometimes simply wrong.'[6]

There were dangerous implications here of a negative culminating moment, a misguided question or conversation, a peak end, or 'a representative moment'[7] on accurate representation of the entire lived experience. Alternatively, acknowledging only that which was observed or spoken during the experience may have provided access to the raw and unfiltered phenomena at play, arguably, the accurate account of experience. However, the mere asking of questions by a researcher disrupts flow, therefore impacting this raw phenomenon. This is something that was counteracted by allowing only free-flowing narratives.

Whilst the experience may be fleeting, the remembering post-encounter is what lasts for people. Instead of deliberating which may be more valuable to a researcher, I decided to correlate information from both the remembering/evaluating self and the experiencing self. If we carry experience with us as Kahneman and Riis suggest, then it was important to recognise what it was that was carried away. It was important to observe what the remembering self believed to be true of the experience, in order to recognise a desire or lack of desire to repeat activity in these circumstances. For example, in Kahneman and Riis' research: 'It was the recalled enjoyment [...] and not the experienced enjoyment that predicted people's desire to repeat the experience,'[8] presumably true in the reverse.

I needed to observe with minimal intrusion and to prepare no questions. I needed to instead document conversations with and amongst informants. I felt that I was faced with several issues that would need to be tackled prior to and within the field. These issues were not compatible with definitive methods:

- Complex ideas were being brought to the fore that may be phenomenological and likely to be relational to *being there.*
- Ineffability and an issue with questions themselves—these essences of *something* would not necessarily emerge via structured questioning or

methods; they may disappear in the questioning or disruption of process potentially brought about by the presence of a researcher.
- Experiences may also be temporal and require a more immediate and flexible capturing of the experience.

I needed methods capable of capturing experience *in the moment* and in reflection. I was also concerned with the effect of definitive methods on the experience of informants:

- Interviews throughout experience may disrupt or alter experience.
- Documentation throughout is an issue with regard to researcher relationships with individuals and groups. This is to do with traditional researcher/ informant structures and the overt presence of a researcher that may alter experience and the recounting of experience.
- The presence of technology poses a potentially detrimental effect on in-the-moment experience. This was perhaps more relevant when working in natural environments within these experiential remits.

These traditionally ethnographic methods and tools may disturb potentially sensitive and subtle experiences, something I wished to avoid.

Within pilot excursions, with these issues floating consistently on the periphery of thought, I attempted to be *in the moment.* I no longer allowed myself to actively consider methods. This was in the hope of discovering a way of *being in the field.* And, I hoped, a way of approaching my informants experience whilst remaining critical.

From this came the unstructured methods of which later became an anti-framework of sorts within the field. These methods aimed to minimise reactivity—to avoid these preconceived issues within the field, I would engage in only free-forming, unstructured conversations and engage in narratives amongst groups that emerged serendipitously. I would take part in all activities that arose, and I would not use technology to record any data on excursions. I would keep discrete handwritten fieldnotes. Additionally, I would only digitally record conversations with individuals and groups as separate to the excursions to reflect on experiences. Having spent considerable time with informants participating alongside activity, these conversations were far more fruitful. Photographic documentation would be minimal and for the most part be used only in order to document the spaces rather than informants. When actively participating, photography would not be an option. Some of the images that you see throughout were taken by informants or practitioners themselves and have been kindly given to me.

Observant Participation and Conversation

I have deliberately referred to my main method as observant participation rather than participant observation since my method involved complete participation in all activities. I observed as I interacted. This is not systemic observation. I was a player, as within participant observation, however, there was a nuanced difference that must be highlighted. This was a more useful way to consider the realities of my fieldwork. I did not stand by and observe; this is an important distinction. However, there is more to this than at first it may seem. The aim was to physically and emotionally involve myself in activities to allow for my own reflection and acknowledgement of subjective experience. This enabled me to build stronger relationships, immerse fully within groups, and experience case studies as an insider. I was physically and socially involved in activities and interactions.

Observant participation allows for a synthesis of 'disparate observations into an integrated whole [and an] arriv[al] at a holistic interpretation.'[9] This is determined by 'my personality and interpersonal dynamic.'[10] This method allows for a subjective and unique way of being *me* as a researcher within the field—one who is influential in data offered and collected within the field. This also affects the serendipitous opportunities for data collection that emerge. According to Moeran, being who I am affects the quality of data (the information) given throughout conversations and more importantly in interaction throughout activities. Though working within the remits of organisational ethnography, Moeran does well to defend my intuition with regard to adopting this term. He states that 'observant participation' is a combination of both emic and etic perspectives—that is, being a subjective participant and being an objective observer.[11] I allowed for objectivity in a highly subjective, bodily, emotional, and involved research strategy.

Within the field and in talking to informants, there is a distinction to be made between what people say that they do and what one witnesses that is actually done.[12] With this comment, Moeran also pays reference to Erving Goffman and his concept of social performance in everyday life. He discusses how observant participation differentiates the access between 'front stage' and 'back stage' and how this method signals 'a shift in role as a fieldwork researcher.'[13] Like Moeran, 'instead of becoming a participant observer in the classic anthropological manner,' I became 'an observant participant.'[14] In doing this, I increased the likelihood and actuality of 'extra mural activities,'[15] that is, being involved in socialising and being considered as a part of the group.

Wilkinson,[16] also sites Goffman's move from front to backstage. She describes the metaphorical 'back stage' as an area inaccessible to an audience (the observer), a space 'beyond the social front.'[17] This is an area made accessible through observant participation. Wilkinson distinguishes observant participation from participant observation as she too found this term unsatisfactory in describing her mode of being in the field with informants since it does 'not capture [her] very embeddedness.'[18] It is vital to acknowledge and 'to know that there is a difference between [informants] front stage and backstage behaviour.'[19]

> [W]e can know the world only because we are part of it [...] no way of being is the only possible one, and that for every way we find, or resolve to take, alternative ways could be taken that would lead in different directions.[20]

Ingold[21] disagrees with the notion that one cannot both participate and observe at the same time or that one cannot be both objective and subjective. He states: '[T]o observe with or from is not to objectify: it is to attend [...]'[22] I hold to this belief. In this chapter, I will demonstrate this involvement with case studies. This will be fully descriptive and lengthy in order to truly capture fieldwork. It gives a more vivid, living picture of the individuals with whom I researched. It gives a more accurate account of these methods, or lack of method, of which I speak and aims to exemplify how themes began to emerge through working in this manner.

The loose ethnographic methods that were used require a little explanation:

- Walking and talking (plus also kayaking and talking, climbing and talking, planting and talking, etc. This will become clearer when I begin discussing case studies).[23]
- Naturally occurring/unstructured conversation recorded digitally and manually, that is, 'unstructured interview.'
- Participation in all activities, that is, observant participation (recorded manually within fieldnotes), not participant observation.
- Attendance at group reflection and sharing sessions; these were not orchestrated by myself but instead by group facilitators. I also attended feedback sessions and discursive group sessions indoors.
- Social interaction, that is, not within natural space but within accommodation, bars, and other locations attended whilst on excursion.[24]

Access and Involvement

Within my first case study, there was a series of excursions to a 'community living' initiative which focuses on working the land, outdoor experience, and spirituality on an island in the Inner Hebrides. Here I volunteered in all activities as all visitors did. Informants travelled here from all over the United Kingdom, Europe, and the United States, and they were of ages from 16 to 60 plus years.

Within my second case study, The Young Adults' Personal Development Project, I worked closely with them as a 'volunteer support worker.' These people come from very varied backgrounds and were dealing with difficult circumstances including behavioural problems, experience within the prison system, issues with mental health, issues with drug and alcohol abuse, developmental disabilities, and learning difficulties. The charity had two projects—one was based in Edinburgh and another in Glasgow and worked with young adults from the surrounding urban area who were identified as 'at risk.' The young adults were aged 18–30 years. The practitioners and volunteers, some of whom were individuals who had been through the programme themselves, were aged between 18 and 40 years.

Within my third case study, I was a volunteer-participant with two branches of a social enterprise that use woodland management, production, and outdoor spaces for wellbeing and sustainability awareness. The first branch is an open access community project that coordinates excursions (The Woodland Weekend Group), and the second is an National Health Service (NHS)-approved The Mental Health Initiative working with a core group of referred individuals. Informants who participated within this project came from Edinburgh, Perth, Glasgow, and Newcastle, England, and were aged between 29 and 70 plus years.

And finally, my fourth case study was a loose community of individuals who met through an environmental sustainability organisation, however who now incorporate the Scottish outdoor landscape into events interested in neo-shamanism, language, the arts, mindfulness, and meditation. Participants within this case study came from Edinburgh, Perth, Glasgow, Linlithgow, and the Scottish Borders. They were aged 29 to 60 plus years.

Happening upon Case Studies

At the beginning of my research process, with only my research questions under my belt, I compiled an extensive folio. This was meant to gather information on all groups within Scotland who may use the landscape as some kind of social, therapeutic, wellbeing-related, or developmental tool. This was a big task considering the unfathomable amount of outdoor and recreational activity within these landscapes in Scotland. Nonetheless, it gave me a clear insight into the extent of activities within this remit and allowed me a clear idea of the practical area. The case studies that became the focus of my research however instead came quite by chance. The Young Adult's Development Project came via a meeting with the project officer of another larger organisation, the John Muir Award, in Edinburgh. She was able to put me in touch with one of the youth development officers at this young adult's charity. The Mental Health Initiative and Woodland Weekend Group happened to take part in a community gardening festival I had been involved with prior to and as I was beginning my research. The Loose Community came after a chance meeting with the group's primary facilitator at what I thought was going to be a seminar (it turned out to be a shamanic vision workshop) and The Community Living Initiative came into my periphery via an invite to visit the community as part of an event organised by the chaplaincy at my university. I could never quite be sure if this was chance or rather a carefully constructed new network of possibilities. Upon reflection, it makes sense that these organisations came into sight; the truth of the matter is that they came whilst I was already immersing myself in networking and navigating this field. Each of these case studies came with entirely different ontological frameworks and this intrigued me. So this is where I began.

Serendipitous Ethnography

Ethnography is 'no longer a jealously guarded "possession" of anthropology.'[25]

On the one hand, within the discipline of anthropology, methods must be flexible. Ethnography is about collecting data in any ethical way that one can whilst in the field. One must engage with one's informant community within their physical and social spaces. It is also key to document one's data in the most fitting manner. This was the loose framework with which I approached my own fieldwork. On the other hand, within further disciplines the methodology is appropriated. In this instance, there is a stronger desire for definitive methods. Within further social disciplines, ethnography may manifest as qualitative surveys, focus groups, interviews and participant observation or perhaps the 'go along' method[26]—most often these are structured research methods with little scope for responsiveness, flexibility, or serendipity when researching amongst the unpredictability of people.

Ethnographic methods are wholly or partly adopted by many social sciences. *My* methodology was, for the most part, anthropological. Navigating this traditionally anthropological, disciplinary rite was enlightening and completely guided by my case study groups and the spaces in which we visited. My methods of gathering data were guided by group structures, agendas, personalities, the weather, and the practicalities of working with groups outdoors. I found that working with multiple groups, with people and dynamics as complex as they are, my field research could follow no strict methodological structure. It is for this reason that I felt the necessity to lay out my methods, at first in a cross-disciplinarily recognisable manner. I problematise this in relation to *being* within the field in actuality. It is due to my straddling of multiple disciplines that I felt this was necessary. Now, however, to elucidate the truly serendipitous nature of ethnography in the field, and why my methods were rather undefined, I will discuss the ethos of serendipitous ethnography within my own research.

> [F]ieldwork is multi-dimensional, requiring ethnographers to weave place, time and context together in ways that make any attempt at defining field-work in methodological terms difficult.[27]

> [The] serendipity component [...] the unanticipated, anomalous and strategic datum which exerts pressure upon the investigator for a new direction of inquiry which extends theory[28]

So long as my research aimed to tell an interpretive and explanatory story rather than a fabricated or fictional one, so long as its narrative would be based on systematically gathered and analysed data, regardless of the flexibility of that system of gathering, and so long as I was basing my research primarily on real-life and face-to-face research with groups of people, I could rest assured throughout the process that what I was doing was valid and valuable ethnographic research. I was going to allow for serendipity, flexibility, and responsiveness to these situations. I did not in any way distort or manage the situations—my job, as I saw it, was to be there, participate, converse and to represent, whilst engaging in a balanced and grounded analysis.[29]

I provide an understanding of the perception and experience of these natural landscapes as shared by several groups—a cultural interpretation from the perspective of the people involved. This aim did not require rigid methods. Instead it required an ability to listen, observe, and interact, so this is all that I did. One of the key factors in this was in building strong and fruitful researcher- informant relationships whilst being sure to acknowledge the group dynamics and developments as they happened. I was by no means unorganised; instead I was comfortable in letting go of plans and research structures, allowing for a truthful submersion within my case study groups and for the activities, conversations, and people to guide my methods. I had the readiness spoken about by Hannerz[30] to 'depart from [my] research plans and research designs' when I ran 'into opportunities that simply should not be missed.'[31]

According to Raymond Madden, ethnography cannot and should not be defined, as the people who do it (the research), and the subjects of study are too messy.[32] Ethnography needs to be a fluid, individual, relational, and group-specific endeavour. Working with groups of people is always going to be dynamic and to carry out successful ethnographic research, it was essential that I formed strong and trusting relationships above all else.

To further support my argument with regard to a loose approach to fieldwork, I address here an essay by Rivioal and Salazar[33] on serendipity within ethnographic research. They believe that serendipity, reflexivity, and openness are not only accepted within ethnographic research but are a strength of its methods. Similarly, Fetterman[34] also argues that good ethnography should not be completely orderly; it should, as Rivioal and Salazar agree, be about 'serendipity, creativity and being in the right place at the right or wrong time,' additionally with 'much hard work and old-fashioned luck.'[35] This is to be based, not simply on a hope, but instead on a 'combination of chance and intuitive reasoning.'[36] The approach should hold narrowly defined research questions, thus allowing for focus, and broadly defined methods to allow for this flexibility and serendipity.[37]

It is not a possibility to concisely detail all of my choices and interactions. Subtle interactions best describe my methods and methodology. Every encounter may be considered as a method of enquiry—less method perhaps, and more a way of being within the field and with people. Data comes from being submersed within the groups which you study and within the spaces and frameworks, and this is where I was. Perhaps this may be another contribution to be made to non-anthropological or transdisciplinary research—a serendipitous approach to researching with people.

Serendipity, Responsiveness, and Flexibility

Allowing for serendipity in research is allowing for happenstance, chance, and letting go of structure. This also means ultimately managing the complex and messy data[38] that you gather. This is not only about allowing what will be to be, but instead developing a responsiveness to what is offered—developing a sense of flexibility to go along with activities and conversations as they develop. It is also about trusting that the data that will emerge is always manageable. I trusted what might emerge in the field and therefore left this concern out of my day to day within the field.

> [F]ield work is no longer considered the stage of data collection, necessary for subsequent analysis, but a place of real interaction and knowledge production [...] the very place where knowledge is produced.[39]

Le Courant[40] is convincing in his statement that fieldwork is no longer merely the data collection stage. As was mentioned when I discussed transdisciplinarity, there seems to be a new research paradigm that has emerged over the past decade. This acknowledged indigenous, interactive, participatory, and reciprocal knowledge production as something with great value.[41] This is something that ethnographers have understood for quite some time. This is spoken about more recently by Ingold[42]—for Ingold, a mutual learning process without discipline, and for Le Courant, 'about the opportunism of the researcher.'[43] Serendipity in ethnography though is more than an openness. It is about leaving behind expectations and focusing on what emerges. For Le Courant, this is the 'interpretive gap.' This is ultimately led by what emerges through observation and is where observation leads theoretically.

This project's transdisciplinary approach considers observation and theoretical engagement as a to-and-fro practice unlike Le Courant. He states that what is observed in the field will determine what he considers through his or her theoretical background.[44] Theoretical background did not lead me to consider through observation, but rather observant participation led me to consider what theoretical frameworks could be engaged with in order to understand what I was witnessing. For me, observant participation led to new theoretical engagements, hence a transdisciplinary research strategy.

Le Courant's notions are not unfamiliar or entirely new within anthropological ethnographic remits. It has become something quite common to consider. However, this is not considered, as commonly, by the term 'serendipity' itself. The 'unexpected,' 'unplanned,' and 'chance encounter' is something that the ethnographer would be familiar with.[45] Davis[46] however uses the term explicitly. She refers to participant observation as a 'serendipitous, chancy, seat of your pants endeavour.'[47] Perhaps then what I suggest is less than radical.

> If you listen to people, instead of asking all the questions, they have a lot to say about themselves.[48]

Davis backs my belief in the value of free-flowing conversation and silence as a method. Davis also claims that her major research method is simply 'being there' as she believes that she is 'the instrument of [her] own research.'[49] Like most ethnographers, she considers her personality, background, and socio-cultural characteristics as influential in her way of being with informants and carrying out her fieldwork. She, like Le Courant, describes making sense within the field through 'intersubjective experience.'[50] She considers the research experience as reciprocal and indeed serendipitous. Davis[51] suggests that this kind of ethnography must be flexible and adaptable, as do I. Another who intimates this mode of 'being' is Ben-Ari[52] who states that within his own fieldwork

> Rich Data was also gathered by simply 'hanging around' when there were no organised activities going on since this allowed me to observe unpredictable occurrences. Along these lines, I found myself listening in to casual conversations [...] a whole plethora of activities that take place at the 'interstices' between more structured events.[53]

Certainly, it is helpful to see these ideas boldly laid out by further ethnographers, somewhat verifying my own intuitions. However, most who have spoken of serendipity in ethnography before me, short of the generic comments on flexibility and openness to chance, seem to focus on the serendipitous nature of their own academic career trajectories. They dwell on the happenstance relationships that have proved fruitful to professional development in academia. These articles and chapters often feel like a lesson in the benefits of knowing the right people within the field or being in the right place at the right time. Davis[54] details her career from a postgraduate degree, via her fieldwork in Newfoundland, to her successful research grant bids. She talks of manoeuvring herself from a young academic to professorship. She does not discuss what I envisage when I speak of a serendipitous ethnography. Similarly, Herzfeld[55] waxes lyrical about friends in high places. His sense of chance ethnographic encounter is entirely based on his career trajectory and professional relationships. Herzfeld travelled through Greece, Rome, and Bangkok. He followed friendships, 'exploring new avenues of research,'[56] explicitly stating that 'friendship both motivates and enables what [he does].'[57] He also speaks of minimal career risks and networks of scholars. A further collection of writings, *Serendipity in Anthropological Research,*[58] presents and explores the notion of ethnographer as a nomad, traversing the world, field site by field site, and argues that this too is serendipitous ethnography. Serendipity for these researchers lies in where ethnographers find themselves due to making various decisions within their career and due to developing interests.[59]

Of course, one may find oneself navigating various paths throughout one's career and across field sites. These paths could emerge serendipitously or instead be based on rational decisions made by individual ethnographers. Serendipitous connections may be made across field sites due to emergent themes. This may entice a change of location or a researcher may simply develop an interest in another informant community. These notions of serendipitous ethnography do not touch upon exactly what I refer to when I say serendipitous ethnography, nor the ethos by which I worked in the field. I have discussed them in order to highlight clear differences between my concept and theirs, and to acknowledge my own contribution to this subject.

Goethean Observation

> [S]triving to observe the things of nature in and of themselves and in their relations to one another [...] with an even and quiet gaze [to] take the measure for knowledge—the data that form the basis for judgement—not out of themselves but out of the circle of what they observe.[60]

> [D]eep clues [...] present themselves to the enquiring mind once it has distanced itself from self-centredness of its everyday relationships to things in the world.[61]

Greverus,[62] like Davis and Le Courant, discusses a mode of ethnographic research that is based on travel encounter, mobile anthropology, multi-sited ethnography, and 'polymorphous engagement.'[63] My practice was indeed mobile and multi-sited, but in terms of being serendipitous it was more than this. Unlike these ethnographers, Greverus also heeds German poet, writer, and scientist Johann Wolfgang von Goethe.

Goethe emerged quite serendipitously within fieldwork due to one informant practising Goethean science and sharing his knowledge with The Loose Community. I realised that, set apart from his literary and philosophical works, he wrote a treatise on scientific botany.[64] His writings here regarded how plant species may be observed and understood as relational to their parts and to one another. Early in fieldwork, I had encountered this mode of observation within mindfulness practice in nature. Goethean observation, for my first research community in the field, was a mode of relating to everything around oneself within the natural environment in a manner that called for a quieting of the ego. This frame of mind, despite being at first only realised when observing this specific group, influenced me in the rest of my fieldwork. It allowed silence to emerge as a method. It allowed me to consider reverie and flow. It subconsciously filtered into my way of observantly participating with informants.

For Goethe, plant formations are observed in relation to the remaining plant kingdom. As an ethnographer, I considered the person in its relation to the other—the environment and the wider social cultural world.

I adopted Goethean observation as an ethic, not towards 'nature' proper as in the physical natural sciences but to social and environmental relationships. Key to this was an acceptance that one must allow for happenstance and observe what was occurring

> As soon as we consider a phenomenon in itself and in relation to others, neither desiring or disliking it, we will in quiet attentiveness be able to form a clear concept of it, its parts, and its relations. The more we expand our considerations and the more we relate phenomena to one another, the more we exercise the gift of observation what lies within us.[65]

The distinct lack of ego, the attentiveness, and consideration of context sound like principles of your average ethnographer. Much to my gratitude, I found Greverus'[66] article. She mentions Goethe's interest in 'deep clues.' Greverus states that a Goethean influenced ethnographic practice would require that we regard things and people in terms of our own context. We would remove ourselves from self-centredness. We would consider the connectedness of everyday phenomena in cooperation with others. A Goethean ethnography would examine all aspects of experience as interrelated and knowable through dialogue.

Where Greverus was interested in Goethe's 'deep clues' and the situation of the travelling anthropologist:

> One stumbles upon it, and it leads one down a trail of clues, stimulating one to see, hear, read and continually pursue new levels of reflection and interpretation until one reaches a text, which is similarly open to interpretation.[67]

And again here:

> If we adopt this stance, not only in our relationships to things, but also to people, including ourselves as observers observed, and if we furthermore expand our horizon to include not just natural but also cultural phenomena, the relevance for considering the situation of the travelling anthropologist becomes clear.[68]

I saw an opportunity reflective of my transdisciplinary ethos. These were opportunities to allow for questions arising, methods changing and adapting to the field, and to theoretical fields expanding. Goethe's principles of relationality highlight parallels in my ethnographic methodology:

> We can see of every phenomena that it is connected to countless others, just as a radiant point of light sends out its rays in all directions [...] be careful enough to examine other bordering phenomena and what follows next.[69]

As Goethe encourages, I saw each insight, individual, observation, cultural or social implication as having the potential to be connected to others. When something—a behaviour, a comment, or interpretation—emerged in the field I was careful to consider what may be bordering this something. What may have come before or fall behind? Rather than focusing solely on methods, questions, and finding answers, I let these things continue to emerge from within interaction.

Serendipitous ethnography is an ethos; it is not a definitive set of methods. The loose methods suggested below followed this ethos by being flexible, responsive, and open.

Being with the Groups: Responsive, Flexible Methods in Context

> Together we seek to enable growth in love, respect and awareness of ourselves, each other, God and the environment. (Community Living Initiative 'mantra')

I will start in the Inner Hebrides with The Community Living Initiative. This initiative is based in an old 'poor house' built from stone gathered from the quarry across the bay and nestled into its curve. Visitors to this initiative were often individuals who knew others within the local spiritual community or were individuals who were invited to return, having visited previously. Sometimes the visitors to this initiative were affiliated with a prominent Christian community nearby, therefore, automatically had access to this space. Many groups however were invited charity groups. They were service users and organisations from low socio-economic areas in Glasgow and Edinburgh. For Allan, the initiative administrator at the time of my first visit, this dichotomy of class and differing uses of the space was sometimes difficult. He moans:

> If middle class people come for respite and the place is full of [charity] groups […] that's not what [this place] is for, it's not for middle class people to come for a holiday!

There is contention here as the space is owned by a community with seemingly differing intentions, to the staff and volunteers that coordinate the space.

Some who visit this initiative are vulnerable, seeking personal development and a space away from their difficult daily circumstances, others were there to find community as 'they see it nowhere else' (Gary).

Allan speaks about some of the young adults who visit the initiative and what the initiative offers that may be different:

> You know, they have got threats of violence or they've got threats of emotional disturbance or they're threatened by sort of social issues, whereas here, you know people are available and want to hear them and want to accept them for who they are rather than what they're expected to be […] it becomes a, a special place for them to be, a place that they regard as their own rather than somewhere they're visiting.

Socialising

One evening, when all others had retired to bed, I was left in the common room with the 'young team.' I watch as Mark, a volunteer leader who was visiting for the summer from Canada, challenged Anthony to scale around the dining room table. He started on top of the table, stomach flat to the surface, and used all of his strength to cling and circumnavigate. He hung off the bottom and winced as his muscles ached. This was taught to the men by a previous female facilitator who was an adept rock climber. She used to do this to improve her strength on days when they couldn't get up the hills due to extreme weather, or in the evening, when people needed entertaining. Omid gave it a shot, and within seconds fell to the floor to raucous laughter. Anthony and Omid are close friends; they met at the initiative a few years ago and have had nights out together in Oban and in Glasgow.

Later, I talked outside with Anthony; it was after midnight and the night was pitch. We couldn't see in front of us, but we could hear the waves. He told me that he was going to have a joint informing me that he was allowed to do this in the evening, though I had my doubts. I had spoken to the coordinator about drugs in the past, and I knew it was forbidden. Not if he wasn't seen, I gathered. Anthony has Attention Deficit Hyperactivity Disorder and smokes 'thirty-five quids' worth a day.' He had 'read up on the drugs he was given as a child' and decided to smoke marijuana instead. It's something that has helped him, alongside this community, to keep out of 'trouble.' He has come here for the summer every year since he was 16 years old; he's 21 now. For Anthony, this initiative provides a sense of structure and security, though vastly different from his daily life in Glasgow, where he cares for his mother and his sister. He told me that he was having difficulties finding a new council home. Currently he lives with his sister, who is now wheelchair bound, due to suffering a stroke in a high-rise flat with no lift. His circumstances at home are difficult and without stability or structure. This is a common theme across personal development initiatives engaged with during fieldwork. For Anthony, this space is a respite from the norm. As opposed to the anti-structure expected within liminal encounter (Turner 1969, 1970, 1974), Anthony has found structure in leaving his everyday circumstances. His anti-structure or subversion of the norm is felt as a safe environment, where daily structures are key. We can see here that through socialising I began to understand Anthony's and others' contexts and motivations for being there.

Being 'Shown Around'

The first time that I visited this space, the week before Christmas, when there were no volunteers or participants and only Allan for company, he took me on a tour of the house. He showed me the common rooms and dormitories. They were stark to say the least. Allan explained, as we opened door after door, seeing one grey stone room after another, bunk bed after bunk bed, that they try to keep the rooms bare, but safe and comfortable. Visitors have no access to electricity so cannot recharge their phones. The agenda behind this is to discourage people from staying in their rooms for too much of the time whilst they are here and so that people get involved in group activities instead. He finds that people accept this. If people have a serious need to call home, they can of course use the landline. Minor, perhaps considered major for some, disruptions of normal behaviours such as not being able to use one's phone characterises this environment. It is these subversions that make up the ethos of this initiative:

> You know we're asking them to eat vegetarian food and sometimes the showers will be cold because we haven't had nearly enough wind to heat up the water and they're living, staying in a dormitory with other people which they're probably not used to and eating with other people and not used to it, so it's a pretty threatening environment really. […] the first day that they're here it's all a bit, 'what's going on?' and then the second day it's like 'I'm finding this quite challenging' and then by the end of the third day they start to settle into it and start feeling more confident and are able to explore and give things another go […] and realising, you know, the amount of freedom and safety [they have] actually in comparison to what they might have back home where, you know, the threats are quite often real. (Allan)

Allan described a situation of ambiguity for these visitors. The scenario for these groups was unfamiliar. I didn't ask to be taken on a wander around the site. This was Allan's suggestion because he needed to do a maintenance check as part of his own weekly routine. By guiding me around the whole of the community's area of the island, Allan was able to reflect on what the space is used for. In doing so, he began to open up areas of enquiry that I was able to probe further. This was to do with the landscape, the sleeping arrangements, and the kinds of groups that journey here and why. I was able to envisage the groups within the bunk beds, within the dorm rooms, the kitchen space, in the bay, and to ask what they may be doing—all in good stead for my return in the spring. I was able to begin to piece interactions together through Allan's memories and associations as we wandered the site. Meanwhile, I was also able to begin to understand Allan's motivations for being there himself and what he *gets* from being there.

Allan had shown me the 'situations room,' a room in which groups do arts and crafts, play games, and reflect. He had pointed out the community's 'mantra,' colourfully painted on the wall, informing me that the word 'god' is a relatively new addition. He insisted that this space, as part of this wider community is not focussed on religion. They do however have five rules of engagement: 'read the Bible every day,' 'account for your time,' 'endeavour to live sustainably,' 'work for peace and justice,' and 'don't spend money unnecessarily.' The first of these I never once witnessed in my time at the initiative. By showing me this painted sign on the wall, Allan is able to tell me more about himself and his relationship with the community. Allan is 'not a religious person,' but he is 'spiritual' he tells me. This conversation suggests that he does not wish to be associated with Christianity of which this wider organisation is so famous. After standing out on the bay together, I press him with regard to the term 'spiritual,' and he says that he 'believes in nature.'

> It's a place where people can come and feel safe, feel held, feel welcome, and can experience simplicity [...] it's a place where people remember that they've been, so it's a place that's special in people's hearts [...] there is a special quality to the west coast, you know the west coast is exposed to the Atlantic and this is where the weather, the weather comes from the west, and so there is a rawness to being on the west coast which is different to being on, in other places, whether in land or on the east coast [...] you are on the edge and I think you can feel that. (Allan)

For Allan, this initiative, most importantly the landscape around it, is a place where people feel safe. He has described the bay to me as somewhere that allows you 'to feel held'; he said this several times. This might be considered as a physical framing. The physical curve of the bay with its surrounding hills adds to this feeling of being held in a protected way—protected from the weather and the exposed nature of the rest of the

Island. However, Allan speaks of this more metaphorically. He described standing and looking out on the bay as being on the edge of the world. He says people remember this place and keep these memories in their heart. This sense of being held, I feel, reflects comments made by Allan with regard to people needing a sense of structure to their lives and to their days here when visiting. Both the landscape and the initiative in this sense allow participants to feel safe, and it may be this that leads to that sense of confidence that Allan sees in the group as they progress.

Doing as the Groups Do

> they remember the experience that they've had here and the experience might be that it's just an adventure because its different and you know without electricity and it's at the end of the track and it's a journey to get here [...] you definitely see a **change**, you know when people come here for the first time then it's a pretty challenging experience, the first challenge is to walk down the track, you know, people aren't used to walking a mile and a half in the countryside in the rain. You do that quite frequently; you know when they're walking down Sauchiehall Street in Glasgow but they don't do it out in the countryside. (Allan)

Groups, as I have done, travelled far to be at this site, and if one was to want to leave, it would be some feat to do so. It is an hour's walk to the nearest bus stop and two hours to the hour-long ferry ride that would take you back to the mainland. The space certainly *feels* remote. As a researcher, I could feel that *sense* of distance. Allan describes it as a threatening environment: referring to stepping out of one's comfort zone, both physically and socially and of also visiting a place with a particular ethos and set of values that may be vastly different from those that visitors are acquainted to. By being here, these people have crossed a mental and physical threshold. They do this by travelling to such a place and by pushing their mental boundaries and becoming open to the group, the initiative, ethos, and to the space. Allan says that after three days, these visitors begin to feel more confident and able to explore. They have the freedom to do this, and he feels they may not go back home. They may not have the freedom to be confident and try new things as he says the threat to safety is quite real—real in comparison to the *perception* of danger in these situations on the island, where they may encounter new landscapes, new people, and new ideals. This is something that seems relevant to the young adults of the personal development project too.

On the surface, it may seem that this organisation is in fact carrying out practical, mundane, and efficient activities with the visiting groups—the activities here tend to be based around chores, working the land, and building structures; yet, the feelings beneath these activities suggest a making special of the mundane. The success of this organisation may be based on these intangible things: made up of mundane and complex elements returning us to the notion of ambiguity. They invite people to experience something different, something challenging and outside of their *normal every day* before making the return journey home.

Structure, Integration, and Sharing

> It's definitely about integrating everybody to live together and to live in community and yeah, we all share the same chores—we do everything together. (Allan)

The volunteers and staff at this community living initiative would say that they are always integrated with the visiting volunteers and groups, though they do sleep in separate accommodation, with heating and electricity. I was fully integrated, I slept in the quarry stone dorms, I did the chores, I did manual tasks—building, planting, and walking back and forth up the track. Chores were shared, as also the outdoor tasks that filled the days. I was considered firstly as a volunteer and secondly, if acknowledged at all despite adequate introductions, as someone doing research. I embraced this. They encouraged a collective rather than solitary experience. The staff were encouraged to 'sit down and chat to people and to hear their story.' This was an established part of the process of being here. It was easy for me to follow suit. Like the staff, who will play table tennis with visitors, start an art project, or play the guitar and games in the communal space, I did similarly. The overall ethos seemed to focus on sharing, both things and experiences. In May, the place was lively with visitors from all over Europe and the United States as well as the United Kingdom. Though still cold and wet, the volunteers and staff would initiate kayaking and walks, swimming in the bay and tours of the nearby salt marshes during downtime. I took part in all of these activities, feeling the freezing water, steep hills, and brisk air. Most importantly I was developing relationships through sharing in these experiences.

Each day at the initiative is begun with a shared breakfast followed by shared chores and facilitated reflection—another serendipitous aspect that has formed an integral part within my methodology across all case studies (sometimes this, within the more pragmatic groups, is dubbed *feedback*). Tasks for the day are set—maybe planting trees, collecting seaweed for the vegetable patches from the bay, or building a structure of some kind. Breakfast and lunch are opened and closed with quirky 'bits'—massages, stretches, and impressions, before returning to our tasks. Dinner is the same, followed by an extended reflection by candle light or camp fire in the team-built roundhouse in the woodlands.

There is the distinct feeling that these things are compulsory. Everyone must not only be there for breakfast, lunch, and dinner but must wait until all are finished and the meal is 'closed' before they may leave. I realised this on my first morning here in May. Nobody woke me on my first morning (I, of course, had no phone and no alarm). I awoke to the women opposite me sitting in their bunks, dressed with hats and boots on: 'It's five minutes until breakfast.' I dragged myself up and attempted to take a minute in the fresh air before heading to the dining room. Victoria, who had been volunteering here for three months came looking for me: 'Can you join us for breakfast, we can't start without you.' Victoria has grown up within the spiritual community associated with this initiative. Perhaps this is the average. For me, as well as participants, this level of structure is different to my every day in the sense that I spend most days, like many others I have encountered, alone and not only do I not eat breakfast, I have no time schedule to adhere to.

The days are structured, but within this structure there is ample time for idleness. I believe it is somewhere between the structure of these days and the anti-structure of idle time that transformations may emerge in these scenarios. It is also where the most distinct themes seemed to emerge. When thinking about transformations, structure and idle time became dominant in my reflexive writing.

The Young Adults' Development Project also facilitates vulnerable urban youths into natural settings where they might challenge themselves. With this group, I worked closely as a 'volunteer support worker' throughout fieldwork.

Integrating with Staff

Andy, one of the team leaders, delivers the 'Journey programme' in Glasgow. When I first met him, he chatted to me about his belief that 'the government may pay more attention to youth unemployment in order to get [the young adults] into the work system.' This project then perhaps becomes more to do with 'creating working adults,' with the desire for enhancing mental wellbeing as a by-product or indeed, an afterthought. Though other facilitators disagree with Andy, it is difficult to dispute the setup of the projects youth programme. The first six months of this programme is based on 'getting the young adults out,' the second six months, 'the community stage,' is focused on CV writing, interview technique, and utilising the new skills harnessed within the previous. This is an environment that intends to, in some sense, 'prepare for work.' With all that that entailed, this was an unfamiliar framework to participants.

The reasons for attending the project by the young adults were varied. The level of independence and perception of themselves as vulnerable were also hugely varied. Where some had difficulty with drugs and alcohol, others were simply unemployed. Some had difficult living situations, some had been in prison for minor offences, whilst others yet had challenging learning difficulties.

Jonathan has autism and only recently began to 'self-travel.' The staff were useful in briefing me on the situations of the young adults. This was helpful when the young adults opened up to me as a newcomer:

> I think mainly the reason for coming anyway was for having something to do. I mean I was in the house 24/7 and all I would do is play on my Xbox and all that and it's not really healthy. So, I thought if I push myself to the limits and then I can find something that I'm good at. (Jonathan)

Attending this development programme is voluntary for all participants. According to Edinburgh support leader, Rory, some people outside of the organisation believed that the young adults attend in order to be able to claim their job-seeker's allowance and other benefits. This is not the case. Not only is this factually inaccurate, the informants also never indicated any similarly minded motivations aside from perhaps the hope to gain employment after attending the programme. The staff do aid participants in issues with housing, support, and psychological or medical appointments. The Edinburgh support leaders, Anna and Rory, agree that none of the participants would be here 'if they weren't looking for something' and that they were looking for 'something' that they 'needed.' For Rory though this need was not necessarily a conscious need. Instead 'something' connects with them when they do arrive at the project or when they have heard about the project. It's an opportunity to do something new and to be part of a group. Again, this is a challenging subversion from their everyday scenarios.

Various Perspectives

Each individual who participated had needs and intentions that were very different. Rory believed that people returned each day, despite difficulties faced, due to the building of relationships with the facilitators and between themselves. Without these relationships, he believed that their 'tentative needs would break down and they would retreat.' This is in line with what some participants told me. Jonathan emphasised: 'There's not really a single activity or a single person or thing that made me come again and again and again. It's just people in general.' When I asked Anna why she thought people kept coming back, considering tensions and challenges she told me:

> I get the impression [that it] is that being part of a group. It's that even though some things are difficult it's difficult for all of us, were all in it together, were all doing it and there's something about being part of something [...] (Anna)

The primary aim of this organisation is self-development. It is the central theme of which the support workers intend to inspire or instigate the participants. Rory believed that some of what the participants may be looking for was simply not achievable 'where they [were] when they come to the course.' By this he meant that the place in which they found themselves, circumstantially, mentally, and physically, was not conducive to moving forward in a positive manner. In his opinion, they needed to develop and to open to developing different sides of themselves. Rory instead described the foundations of the course to be like 're-parenting.' The young people have basic life skills; however, they were attending because they were unsure of 'what to do or where to go next in their lives.' They no longer wanted to continue as they were—drinking, 'smoking pot,' or fighting. According to Rory, often 'something, perhaps only one thing didn't work out' for these young people. His intention was to re-energise them, 'to go out and find out what they want to do.' However, he insisted that 'they've got to go out and figure it out for themselves.'

> I think the course, hopefully the course, enables that for them or creates the context for them to do that but it's very much driven by them, we can't do that, we can only provide the context I think and you know kind of prod them and yeah, it's up to them. (Rory)

Anna is the first person that potential participants will meet. She visits schools, pupil referral units, and social care facilities. She speaks with social workers and other public sector facilities dealing with unemployed, at risk, and vulnerable young people. Anna is the person who communicates with them what the course is about, and she told me that it is a hard thing to describe. She described it as a personal development programme:

> An opportunity to work on communication skills and team building skills [...] it's definitely, part of the communication right at the beginning of the course about what the kinds of things that you're able to achieve when you're on the course, if that's what you want to work on [...] and building confidence, and confidence seems to be the one that everybody comes back to for me.

Combining Research with Other Roles

The distinction between being a researcher and a support worker was often blurred by the group facilitators and the 'young people.' An excursion to 'the bothy' was an example of this. I had been asked to attend as one of only two female support workers, meaning that I was a compulsory member of the support team. This left my role somewhat ambiguous, sleeping in a room with only one female participant and one psychologist. Often, I was asked to witness medication taking, prevent bad behaviour, and attend staff meetings to comment on individuals and how they were dealing with the five-day trip.

I was presumed a support worker by both staff and participants. Initially uncomfortable, this became key in the rapport that I had with this group. I was still able to build and maintain a close and equal relationship with the young adults throughout my nine months with them. They often referred to me as 'cool' and 'not one of the adults' or if someone was doing something that they shouldn't have been Amy often said, 'och Becky doesnae care, she's sound' should anyone have shown concern that I was standing with them. Being considered as a person of authority despite my behaviours was helpful in building trust with my informants. I was not a threat to them. In fact, I was often offered Sam and Amy's joint as they left the base (I declined politely). Sam considered himself to be dependent on drugs and alcohol. Amy is a single young mother who has issues with anger management, anxiety, and depression.

Being of a similar age as the majority of participants, becoming part of this group, for me, was perhaps easier than if I had been significantly older. Having grown up in Edinburgh helped too. All the participants were vulnerable to some degree. It was a difficult balance to maintain, having the trust of both the facilitators and the young adults as well as ensuring the safety of the 'young people.' It required some care to be accepted as not a threat, nor another support worker who would challenge them. This was something I was forced to navigate, to degrees of success.

Building Relationships

As a rule, throughout fieldwork, I would not attempt a lengthy or personal conversation with an informant until I had been within the group long enough to have developed a significant bond with each them. This bond needed to be one that would allow for openness and honesty. This resulted in conversations that flowed quite naturally and saw both researcher and informant on an equal footing. I spoke *with* my groups rather than in a structured, researcher-researched manner.

For several of the participants, the face-to-face contact with people from this group was the only contact that they had throughout the week. Getting to the point of open and lengthy conversation was an achievement for both them and myself as a researcher.

> Everyone I know, I have a laugh with and everything but it's not like face to face if you know what I mean? It's over social media or it's on a headset talking, Skype everything like that, you know people that I've met all over the world which is, you know I've got people from America I talk to that I get on very well with and everything's the exact same [...] Do they know I've got autism? No. But, at the end of the day, you're getting on with people that perhaps you only met for the first time [...] if I had to meet they people face to face, I think it would be a lot harder and challenging than doing it over a headset because they don't know you [...] They don't know your facial features or anything like that. [...] I feel a lot more comfortable [...] rather than face to face because me personally, I really don't like being myself. (Jonathan)

For Jonathan, attending these excursions and meetings is one of the first times in his adult life that he has presented himself to a new group of people outside of virtual reality. This was the first time that he had voluntarily done so with the intention of trying new things and being accepted. Jonathan also attends a youth club. These two spaces are his new contexts of social interaction. For Jonathan entering into a social situation and being initiated into a social group is ambiguous. He questioned whether he would be welcomed.

Being within a group of peers was a different situation than some were used to. Further, being within a group where behaviours were monitored or challenged was something that no one within the group was used to. All participants had either left school early or did not attend school—the average situation for this to occur. The opportunity to forge relationships in this environment was novel. Though behaviours were monitored, encounters often erupted and escalated within the group. This was both out on excursion and at the base hub of the project. This was often referred to as the group 'storming' by facilitators Rory and John. This referenced educational psychologist Bruce Tuckman's model of group developmental stages (1965). 'Storming' was believed to precede the 'forming' of the group and be followed by 'norming' and 'performing.' This invariably set the tone for much activity and the programme in general. In fact, many of the activities carried out within this group paid reference to group developmental theory. Tuckman's model may be compared to the stages of liminality, which I will discuss later in Chap. 5.

Walking with Informants

I walked with this group, the first of their several hill walking trips in the Pentland Hills. Throughout this walk, we stopped on several occasions to bring the group back together to look at the map and discuss where we might be in relation to landmarks. The group was split, with several keen on listening, observing, and attempting the 'challenge,' whilst others desperately wanted to simply keep walking.

We stopped for lunch and sat in the snow and on our bags. I was sitting and standing, sitting and standing, unsettled due to the cold. There seemed in general to be quite a bit of standing in the cold, waiting for others, sussing out the pace of the group, which frustrated Amy immensely:

> Every time we stop I just can't be bothered anymore, when we're walking, I can be bothered.

Amy was frustrated to the point that at the end of the day she didn't want to participate at all anymore, marching forward saying: 'F**k this, I can walk back to the minibus masel.' This waiting and walking gave me a clear insight into not only how the group adapted to being in these environments but to the evident lack of experience of situations such as this. It also provided some insight into the 'storming' of which the team leaders had warned me about.

On a later trip, further in the 'progress' of this group, I saw that the size of the group and the differing needs, abilities, and attitudes affected the walk itself. Prior to the five-day bothy trip, Anna had asked Amy how she felt about the upcoming excursion, to which she responded: 'F**k that! It's sore [...] you know how long it took me to get up that hill?' Here we might be seeing apprehension due to an individual's perception of their own ability and pace. The way that, considered by this group, extended walking impacted the inexperienced body was detrimental to the preconception of future excursions. Pain is not pleasurable to this group. When group leader Rory asked the group to reflect upon their first bothy experience, he asked the group to consider whether physical exertion and tiredness or aching muscles may be considered as a positive outcome. The consensus of the group was categorically no, no it may not. Conversely, with experienced walkers, I often found that this aching and tiredness due to physical exertion outdoors was something that they looked forward to. It was certainly something that was commented on frequently as a good feeling. For the young adults who lacked this contextual experience, it was not considered so.

The group was provided with kit at the beginning of these excursions. They were given waterproof clothing, boots, backpacks, hats, and gloves. Being involved in the organisation of kit and bag packing prior to excursions also gave me a useful insight into the kinds of activities that were built into making these excursions a 'challenge.' It also enabled me to recognise this as a potential framing activity. The kit must be brought along regardless of the weather. This is an exercise in being prepared for anything—perhaps a metaphor for the life skills which the organisation aims to instil. On one occasion, Charlie, the Glasgow group's programme manager was assisting in the Edinburgh group's semi-final excursion to the Bothy. He dropped us off at Glencoe. We were 'dumped' at the foot of Buchaille Etive Mor and Buchaille Etive Beg and were to walk the Lairig Gartain. The group headed off, and minutes into the walk, it began to rain. Myself and another volunteer support worker stopped to unpack bags in search of waterproofs. These were communal bags being carried by two members of the group; the rest had been taken ahead to the bothy in the minibus. We helped Guy to tie his shoes and to put on his waterproof trousers. Guy is German and has autism. He spoke English well but struggled to communicate with the group. He was often teased and the butt of the group's jokes and gripes. On this occasion, he was perceived to be keeping the group behind. It took almost three hours to walk to the river bank before which we encountered steep terrain, streams to jump, and rocks to scramble over. The group led one another, waiting, sitting, slowing down for others, and attempting to be patient of each other's capabilities. We needed to cross the River Etive with only one canoe, 15 backpacks, 15 people, and food and drink for five days as well as a fridge. By the time we were beginning to cross, we had been walking for over five hours, and the group was exhausted, angry, and some, hostile.

Ingold[70] describes walking as knowledge making. This is akin to ideas regarding peripatetic movement and the romantic walkers of the eighteenth century, when walking was considered as a 'fundamental expression of agency.'[71] I believe this is an important concept to note when considering this case study group in particular. For all but three of the participants in The Young Adults' Development Project, most had no context in which to place this lengthy encounter with walking as they had no previous experience of such a task. Two of those that did, Amy and Sam, understood these experiences as part of further youth development projects they had been a part of. Another, Keith, knew of walking only as something structured and goal orientated as his experience was in the cadets when he was younger. I agree with Ingold[72] in the sense that *something* happens with regard to the group's knowledge of the space and themselves throughout this walking from A to B. I also recognise a shift in their understandings of their own agency. I am not sure however if these informants were aware of this. They did not speak of it in such a way. However, the agenda of the facilitators suggest there may be something in this. Whether this knowledge is to do with one's own capabilities, cooperation, or affirming one's own agency depends on each individual's perception. The ability of the rhythm, pace, and temporality, indeed reverie—of which Ingold (Ibid.) hints at as transformative—was not always evident within this group. For the facilitators, a steadfast knowledge of an individual's progress or transformation was not possible unless articulated:

> Paul for example is talking so much more, so obviously all that we can work on is our outward signals that would suggest to us that he might be gaining confidence. I can't tell you whether somebody has gained confidence or not. That's within them but what I can see is him talking more and him talking in front of people more [...] (Anna)

Observing Physicality

In addition to a group activity being influenced by the constructed ideas and frameworks of a facilitator or organisation, individual group members affected the dynamics of the group—their mental state, thought processes, cognitive abilities, behaviours, beliefs, apprehensions, fears, strengths, and so on all contributed to this dynamics. As well as this, each individual's physicality (or ability) affected the pace and the motions of the group. When some slowed down, others also slowed down to remain nearby, whilst others walked ahead. Their bodies twisted and looked back or they looked forward to another in the fore. Some bowed their heads, methodically watching the person in front's footsteps. The non-human—what was seen, touched, and walked over—affected this movement. The space, objectives, and ability affected pace, the pace then affected awareness of surroundings and subsequently the bodily and emotional experience. In another of my case studies, this is the objective—to slow the pace in order to allow for an altered, subverted, experience.

The Loose Community bases their journey through the landscape on being mindful and focusing or attuning to the landscape. The young adult's walks, due to the agenda of the organisation, instead are steeped in personal challenge, team cooperation, and communication. In this instance 'slowing down,' as opposed to 'being mindful,' is undesirable. Here it simply amounts to 'waiting for the slow people' (Amy). The difference between these case study groups may be in point of reference, or indeed a preconceived ideological framework. The young adults have little reference, or experience of, not only the landscapes but also what they are supposed to offer. I consider Senda-Cook's[73] work on experiential degradation here. By not being aware of 'the code,' or supposed 'appropriate behaviours'[74] associated with these landscapes, might the young adults' experience be considered as lesser or at the very least different? Are they 'real recreator[s]'?[75] The organisation of which they are part comes with their own ideological and socio-political framework: this framework is based on development, challenge, and support. It is based on inspiring individuals to change their lives and to contribute to society. The Loose Community's framework, on the other hand, is related to the inner self, the group, and their own relationship to the wider non-human world. These encounters do not come with the pressures that seemed to be attached to The Young Adults' Development Project. This noticeably impacted the kinds of activities, pace, and accepted physicality within excursions.

The group's understood intention for the journey, due often to the group facilitators' intention, influenced the way that the group felt they were to experience the landscape. This also affected the way that the group subsequently walked, including the pace at which they walked. This affected what they saw (or the time allowed for observing), and subsequently the experience itself. This had implications whilst travelling through the landscape. There were implications on the overall awareness of the landscape and awareness of one another. These variables influenced the way in which the journey was approached by individuals and indeed the way that they may be experienced. It may be the case that by being told that one must 'challenge' oneself in travelling from A to B, a group may have a negative view of the experience from the get go. Sitting outside the bothy, I chat to Amy. I'm seated on a stone whilst she lies in a hammock with Sam, between two trees:

R: You told me that you hate being outside and you hate walking, and then I see you hiking on up the side of that mountain and sitting on the rock with Sam three times today! What's that all about?

A: (Laughter) Aye I know. I dinnae like walking, I'm a lazy c**t […] (Laughter)

R: I think that you just don't like being told what to do!?

A: (Laughter) […] That's true, I really dinnae like being told what to do.

When left to their own devices, I witness Amy, Sam, Keith, and Guy each take several walks for extended periods of time around and further afield from the bothy. This is strange for those who tell me that they don't like to walk. There is a correlation between disliking activities and not having the freedom to do as one pleases. Interestingly, these walks are slow and meandering, allowing time to sit, chat, and contemplate. Charlie had said to me the day before this conversation with Amy that, if given downtime, the group will become bored and become 'childlike.' They will look for things to do and begin to settle in and be happy to 'just be.' Charlie was speaking of idleness. The space and time to be idle, for Charlie, is as important as the planned activities of the facilitators on excursions. It is idleness that leads to a more creative use of time, solitary time to reflect, and time to engage with the group in an unstructured manner.

Exploring Contradiction

> I think, I think I feel better in myself than I do on the outside world if you know what I mean [...] I think inside [taking part in this project] these six months has been really challenging and it's been really amazing and it's good to sit up, like up the stairs in the bothy by yourself and just sit there thinking, well, you're feeling good in yourself. You're thinking, well, I'm doing this, would I say the exact same six months ago? No. It's been a really good experience [...] I've enjoyed every second of it. It's been really good and I wouldn't change this for the world, I wouldn't. Like if six months ago, I got told that this is what you're gonna end up doing—you're gonna end up having a laugh with everyone and all that, would I agree with it? No. (Jonathan)

Within the same conversation, Jonathan provides a slightly contradictory perception to the one he has given above. Where above he refers to taking time out on his own to reflect on his experience and to enjoying doing so, below he details how he dislikes being on his own:

> See I don't really like doing stuff on my own, as in like walking about on my own or anything like that like, you know what I mean? I'd rather have

> people that I know and people who are there you know what I mean? rather than being on my own like, within the first couple of weeks like, when I was learning to self-travel I used to hate being on the bus on my own. One, it was boring and two, I never had anyone to talk to […]

Contradictory statements are not unusual for this group. In fact, contradiction was a mainstay across all case studies. Often, when explored, this was where my most interesting findings lay. Many in this group demonstrate a difficulty in thinking about and articulating their experience and in relating these experiences to themselves and to their everyday life. It is conversations such as this that began my pondering on the relevance of abstraction within these experiences.

In Jonathan's case, it may be that he was referring to how he felt at the beginning of the programme (he refers directly to his experience within the first two weeks). He has not acknowledged that there was a difference between these feelings and the more recent feelings of enjoying solitary time later in the programme. He says that he wouldn't have thought it possible to have done such a thing six months previously. At the beginning of the programme, he disliked being alone. Yet, he has also described how he was not comfortable in group situations and rarely had social experiences outside of virtual reality, video gaming, and social media. One might gather from this that Jonathan is seeking out group situations. Having experienced this here, he may be expressing, subconsciously a desire to continue with these kinds of experiences whilst also enjoying being himself and unchallenged on his own. This was something that he was used to in his everyday life. This time however, he is in a different circumstance where he could consider his self in relation to a group dynamics.

Volunteering

I was a volunteer-participant in two branches of one of my case studies. This was a social enterprise. Both branches use woodland management, production, and outdoor spaces for wellbeing and sustainability awareness. The first branch, The Woodland Weekend Group, is an open access community project that coordinates excursions for the public. The second is an NHS-approved The Mental Health Initiative working with a core group of referred individuals. The public branch has a focus on forest management and conservation. They clear invasive and exotic species of plants and trees, build wooden and stone structures, plant trees, dig ditches for irrigation, and liaise with the community leaders and local forestry workers. Within The Mental Health Initiative, the work also includes hand-made crafts, woodwork, and food production. Though the work appears to have an interest primarily in sustainability and conservation, this is rarely the agenda or motivation for attending as described by participants.

Informants, particularly within The Woodland Weekend Group speak of meeting new people, getting out, doing something different, feeling a sense of pride, getting exercise, and doing something that they feel fits in with who they are or because the group has become like a family (among many other reasons not to do with conservation or sustainability) as reasons for attending excursions. Those who attend The Mental Health Initiative are unanimously there to 'feel better.' For this group, these activities are explicitly intended to improve one's wellbeing. This may be found in the sharing of values such as within the sustainability agenda. Similarly, to the young adults, these participants suggest a somewhat different agenda to the organisation's facilitators.

This organisation has gone through a period of 'confusing identity' according to Ross, the organisation's founder. They have been facilitating events, courses, and workshops engaging in mental health, corporate team building, volunteering, sustainability, and carbon emissions. For Ross, it is

> [...] about environmental sustainability and getting people to engage with their value systems and to bring about a change by getting people connected with their environment. (Ross)

In the past, Ross has struggled to explain the organisations' work. Since then, having sought marketing advice, the organisation is now based 'around people and around planet.' The charity concerns itself with the welfare of people and the welfare of the planet and each branch roughly engages with these two themes.

According to Ross, the organisation deliberately uses phrases such as 'adventurous weekends' and encourage with words such as 'have fun […] come join in with nice people you know and do some great activities together […] have a sociable time' within their promotional material. He is very aware that regardless of the further aims of the organisation, group motivations are swayed more so by the social aspects in both the public and private branches of the groups. There is a definite belief in the psychological benefits of both the weekends and the base hub activity. They are simply not emphasised to the same extent. Whilst originally social aspects may have been a by-product of their sustainable agendas, it seems that this by-product is now foregrounded with the sustainable agenda appearing as a means to this end.

> I think that different people come into it for different reasons so their motivation for engaging is different but the benefits are the same […] people might not engage with a project if they thought that it was about mental health because their primary interest might be in environmental sustainability or exercise or wanting to be out and listening to birds in the woods and then it's just a by-product. I think a lot of people if you ask them they'd say that oh yeah there is that benefit too but it's not the primary driver that makes them decide to come to something. I think with the mental health side of things it's the inverse a little bit because the primary driver is they want to feel better and they believe that this will make them feel better or they think it's worth trying or somebody else has told them they should try it and they give it a shot and then the connection and the sustainability and the joy of the environment is the secondary thing to that […] (Ross)

Ross suggests that the original agenda of the group may not in fact be the reason for continuing to attend. Instead, it may be people's desire for the other social aspects, and the desire to be within natural spaces that encourage further attendance. Ross still states that people in the group may have sustainability as their primary agenda; however, this was never articulated to me. This may simply be the agenda of Ross and some of the facilitators. Something can be said about participants wanting to be a part of this considered virtuous intention, not uncommon in the endeavour for feeling positive about the self. As there are people who desire these social experiences within nature, these activities are popular. Ross suggests that how the woodland weekends are pitched would invariably alter the current dynamics. Should activities be advertised as an initiative for wellbeing or mental health only, they may not be as well attended. The environmental sustainability pitch presents the activities as something that may entail those 'known' by-products stated as reasons for attending without being explicitly thought of as a mental health project.

These 'known' by-products would only be recognised by those with previous context of being outdoors or engaging with groups and physical work outdoors. The Mental Health Initiative participants may not attend something described only as an environmental sustainability organisation as it does not specifically outline the driving forces for their attending—to feel better, recover, or improve one's wellbeing. This is a complexity that fascinated me. The groups seemed to be doing the same things; however, due to how the activity was set up, the activities were perceived as having differing capabilities due to people's explicit intentions.

Being Idle with the Groups

For Charlie, within The Young Adults' Personal Development Charity the freedom to 'just be,' free from the stresses the young adults endure in everyday life is where transformed ways of thinking and changed behaviours can be seen. At the very least the young people are forced to 'think of something other than playing with technology.' According to geographer Chris Philo,[76] 'somewhere like the home is not conducive to such reverie, being too dominated by the cares and activities of adult life [...]'[77]

In the image above, some of the participants play various games outside the bothy, and Sam and Keith build a shelter out of scraps that they have found in the tool shed and around the area. Idleness was something I encountered within all of my case studies, not least The Mental Health Initiative;

> during the spoon carving [...] there's no set time to do it and I think that, I find is quite interesting because you can take however long you know you want with it [...] that is the sort of rhythm that seems to, to kind of set itself and, I've found that at times I was wanting to slow down a little bit and I would start to listen to Greig going on about his [laughter] whatever

> he was talking about and there was other people chipping in and at times I was in a little bit of a dream land [...] (Arnaud)

Idleness, for French philosopher Jean Jacques Rousseau,[78] is at the core of moments of reverie such as the ones described by Charlie and Arnaud. Whilst Rousseau (Ibid.) defined reverie as the act of being idle or 'doing nothing much,'[79] I might argue here about the level of inactivity necessary to be considered as idle. When I speak of reverie, I speak of action that removes individuals from perceived adult responsibility. As Philo suggests:

> an imaginative revisiting of the reveries—the absent-minded daydreaming—of bored and 'idle' children.[80]

The aim of simply 'being' in the space is something that is spoken about often within the case study groups though they certainly didn't 'do nothing much.' Their activity is tactile and active, however, removed from the 'adult responsibilities' or everyday life of the majority of participants. This often allowed for a state of flow. Whether lopping and pruning in the forest or crafting with natural materials, these case study groups exhibit tendencies towards relaxation, reverie, and contemplation within these activities.

Tactility

Within The Mental Health Initiative, the favoured activity is woodworking. The group also spends one of their three days a week in the woodlands and one contributing to the maintenance of the peri-urban site. One full day a week is dedicated to individual woodworking projects. The charity is self-sustaining in this way, and each of the branches contributes to the other:

> Wood is just such a great, you know tactile material to work with, so whether it's willows or whether it's you know split hazel or whether it's bigger pieces of wood or you know, really big pieces of wood, you know you can eh... you know wood is a really great material to work with cos you know everybody can sort of pick it up, nail it, screw it, you know, glue it,

> carve it, chop it, you know, turn it [laughter] it's such a kind of great materiel as compared to metal or plastic or other alternatives and it is a material which is very, very natural. We sometimes forget that every piece of timber has been a tree at some time. Sometimes you don't make that connection [...] so we're kind of able to work with both sides of it [...] we work with the trees while they're living and then we work with the timber which comes from the trees which are cut down through management, so we kind of get two bites of the cherry... (Patrick)

The *being* within a space that is different from one's normal reality is part of the process for these case study groups, particularly for The Mental Health Initiative. For these informants, it was in making things, perhaps labouring, that they found their sense of reverie and simply *being*. This was uncomfortable initially for some within this group, but something that became a vital part of their recovery. This was to do with not only simply *being* in the space but being there surrounded by others, others that were initially strangers. For founder Ross, participants liked doing something with their hands. Some participants liked building, making steps and hand rails for example, and being constructive and others,

> just love being out in the woods doing pruning you know [...] they really get a lot out of that [...] and then some, I think like the socialness of the group. (Ross)

Gaston Bachelard[81] preferred the notion of 'conscious imagination' when considering the state of reverie. This perhaps describes more fittingly the state that my informants found themselves in within The Mental Health Initiative. They were within thought whilst engaged actively in imaginatively transforming a felled tree into spoons, spurtles and other objects. What appears to be key is not only doing something constructive, imaginative, and contemplative, but doing it in the company of others:

> when you do things on your own guess what you do? you sit and ruminate and think, even when you're in the bloody swimming pool and the last thing somebody like me needs is any more time thinking. trying to think myself better [...]

> [...] sitting in a collective in terms of doing it, it was just a bit of quiet time and when that happens that's good but that's different than sitting in my own back garden you know smoking a cigarette and going you know, how do I get better? how do I get better? even if I was whittling a spoon at home it's not the same. (Greig)

Being Immersed

The physical senses play a key part in apprehending the natural environment, not only for 'receiving data from external stimuli but as an integrated sensorium.'[82] This understands sense qualities as integral to perceptual awareness, not simply psychological or neurological but key to the engagement of the conscious body within the natural environment.[83] For Berleant, attempting to explain this sensual awareness, perceptions, understanding of our self, and aesthetic experiences in circumstances such as these is difficult.

Experiences are consistently affected or effected by our presuppositions, habits, and traditions.[84] We must then attempt to understand the contexts in which these activities took place—both the informants' ways of learning, doing, and being, as well as the organisations and groups that they are part of.

Data began to suggest that our relationship with the landscape was not merely a social or a cultural one but a natural one—psychological (in the mind) and phenomenological (within the body and mind) for some, this was felt as 'innate.' They believed that their relationship with nature was innately human. Though sociality, susceptibility to cultural influence, and cognitive processes are innate, something else began to emerge. It became apparent that perhaps the innately positive experience of nature may be considered, or at least that 'nature' may be subconsciously 'familiar' to humans regardless of experience. Comments suggested that we may need to question whether modern culture would have anything to do with it:

> I think it is kind of out with people's every day experience and yet it's something like our ancestors did a lot of you know, our ancestors had to kind of cut down trees and chop them up to keep themselves warm to survive you

> know so it's something that's really kind of deeply embedded within us I think and just like sitting watching a fire is something that's kind of hypnotic, mesmerising really and I think that's also because you know our predecessors would have just done that all the time, it would have just been so essential a part of their lives so whenever we get the chance now to chop down wood and put it into a fire and watch the fire well your kind of connecting with that, with your ancestors and I think that's part of the satisfaction that it generates. (Patrick)

> I was wondering, I mean we are part of the natural world and I think that there is just a natural connection with nature which I dunno if it's something that... I think its mutual, there's just something that, it's not coming just from me thinking I'm just going to go in nature because that's what, I think it's a mutual thing. There's sort of a getting in tune with yourself when you're in nature. For me it's something that relaxes me. [...] yeah, I guess both in tune with myself and in tune with nature because I... I was just wondering why that was? and I think that for me that's because we are part of the natural world so it just comes naturally, just like the rhythm of the waves in the sea is meant to actually connect with your brain and relax the brain and it's just a natural thing that we feel. (Arnaud)

To suggest that our relationships with or our experience within natural space is in any way innate may be considered as risky. If these interactions are not only social or cultural but natural as well, as an ethnographer I was faced with a new challenge. I could no longer only describe and analyse what was said and seen, this would not give an accurate representation of what in fact I believed to be going on here. When approaching such field-emergent themes one must turn to further disciplines, from where according to Maurice Bloch[85] traditional anthropologists may have shied away for fear of being deemed reductionist.

According to Bloch,[86] ethnographers must acknowledge the unreachable cognitive process. If we merely analyse what is seen and said, we may miss the crux of the experience. This was specifically important if I was thinking in relation to human experience of the natural world. When we speak of the 'natural' world as being beneficial, transformative, transcendental, and as having an effect on mental health, we must acknowledge that there is more to it than walking in fresh air. There is something beneath that may be unspoken or ineffable or indeed stated as difficult to explain, that only those disciplines which engage with the psychological and physiological may be able to tell us anything about. It was imperative then to not only have a secondary research strategy but to immerse myself in order to be able to reflect on my own physicality within the space and my own emotions.

Taking Part

At the base site of The Mental Health Initiative, we sat beneath a tarpaulin, stretched from roof to fence, next to a hand-built roman roundhouse made of pillars of tree and draped with bracken. On one occasion, Jess was using a traditional hand-built wooden lathe to carve a spurtle[87] or a dibber.[88] Jess had finished 18 months of therapy with a psychologist and was back seeing a psychiatrist again when she began with this initiative. She had been telling herself 'for years' that she was going to volunteer and had looked into another organisation but hadn't felt ready to commit previously. Her psychiatrist had referred her here. Nathan joined Jess to work on the lathe, whilst Holly worked close by in the sunshine. Holly came to the project differently in that she was in a period of recovery. She had 'been through the worst of what was going on' and had 'started to get some clarity on what she wanted to do next.' Both women were in a state of uncertainty and seeking 'something.' Originally Holly had attended the public reskilling events that the organisation hosted and had met the facilitator Josh. She was not referred as the others were until she asked her doctor directly:

> Just because I had remembered that I had previously done some woodworking and really loved it and had, through my process of, I suppose trying to improve my health and trying to understand myself and what is it that I need? Like, what is it that makes me feel like I can cope better with whatever challenges I may face? and I pinpointed that creativity is really important, but not just general creativity, not just anything but what was missing was the doing something with my hands. (Holly)

Facilitator Martha and I, alongside Fiona, worked underneath the tarpaulin, whilst Greig split firewood at the back, bashing the hand-made wooden mallet thumping down on the axe that was laid indenting the wood. Grieg was lucky in that when he was 'right at the end of his tether,' he had an appointment with his psychiatrist. He pleaded with his doctor for something that would 'give his life some structure,' and his doctor had suggested this organisation: 'I didn't have a clue, he told me to go away and google it myself.' With the help of his ex-community psychiatric nurse, who seconded his referral, he was signed up to join the group. We chopped away for a good hour or so with others coming over to comment on my progress. Holly gave me a hand with the hook knife, and she taught me how to hold it and how to protect my thumbs. I slipped within minutes and cut through my gloves and finger, and blood seeped through. She had her own hook knife and showed me how you might wrap something around the blade where your thumb rests to prevent cutting yourself. She told me that it's all a matter of practice and feeling what's right. She held it close to her chest, thumbs overlapping and contorted it into different positions that may be suitable. This ethos of skill sharing is something that is referred to almost daily within this group and something that I would not have discovered if it weren't for taking part in this activity.

Learning and Sharing Skills

> It's actually quite a lot for me about the learning environment so I feel very much like, I'm learning a lot obviously about the things were doing, like specific things but I'm also learning a lot about myself, like remembering or re-learning things about myself in terms of who am I in a group? you know what are the dynamics? [...] this is a really supportive space and we also have people who are showing us what to do and are supporting us and taking care of us [...] I don't perceive any pressure so I'm able to get into a state of effortless learning and feel my own successes [...] (Holly)

> I think with natural materials, maybe that you cannot just make it bend to your will [...] It's not just, you have a vision and you just make it and bend everything like with a very Victorian industrial kind of perspective on things. If you're working with wood or willow or what have you, you have to work with what you can find and it maybe won't be exactly what you were looking for and you have to have that flexibility of approach and of expectation of outcomes [...] (Ross)

There is a metaphor within Ross' comment. Working with wood asks that the maker be patient and flexible; he likens this to the reality of all of us, particularly those within his projects who do suffer with mental health issues.

> If I was to say I'd find one more stressful I'd say Tuesdays because once I'd made one good spoon I wanted to be really good at it. So yeah, I've kind of stopped putting pressure on myself but for a while I was 'I wanna get really good at this really quickly' and that's pointless. (Jess)

Jess is talking about the pressure to keep the activities that she devotes time to untainted by what she perceives as failure. She mentions going to college and the inevitability of having to take a day off. For her, this would then mean that 'college was ruined,' that she was no longer doing well due to 'one bad day,' and this made her feel 'out of control.' She would then subsequently feel a loss of control over several things in her life. I mention this here as I believe this has a lot to do with what Ross was discussing with regard to working with natural materials and being flexible. Jess was not able to be patient or flexible with herself and felt unbearable pressure for things to 'go right' in her life. She echoes this when she says that Tuesdays (the woodworking days at The Mental Health Initiative) were originally stressful for her as, after realising she was a talented spoon carver early on, she had felt pressure to continue and to not make mistakes. Wood carving doesn't allow for this kind of mentality. Due to the nature of working with wood, it snaps and breaks and doesn't bend to force, it requires patience and an acceptance that things may not go to plan. By doing these activities, participants began to be more accepting of mistakes, and this led them to be a little more patient with themselves:

> I think some of the guys, not now so much but maybe in the early days, there was a lot of avoidance about finishing things in case it was done and then it would be judged and what if it wasn't good enough? but I think it's, If we can help people to go through that process and develop those skills and to finish something and to stand up and be counted on it and it actually might not be very good and to be OK with that and to believe that it can incrementally get better… I think that's really important. (Ross)

Making a wooden object is a reflective process. One starts with a stump of wood, sawed off a felled tree. Between this object and a crafted spoon there lies not only a physical but a mental challenge as well as the space to ruminate in flow. Whilst making with informants, and whilst relaxed in a sense of flow, we were able to converse in an unstructured way without force—equal in our mindful, tactile engagement and conversation.

One afternoon, whilst teaching me how to work with a hatchet, Fay, the administrator of the organisation and support worker within The Mental Health Initiative, insisted that what is useful is that for most of the participants woodworking is something that they haven't done before: 'It's not like baking or art.' It is not familiar to the participants. This gives them the freedom to experiment without fear. It allows them the opportunity to attempt something new without fear of being upset or embarrassed should it go wrong. For Fay, this is important as 'we don't get to do that as adults. It's always, I can't do that or I've never done that.' Therefore, people in general shy away from pushing their own boundaries and being creative. The work that the facilitators carry out here gently pushes people out of their comfort zones. I have seen that some people take it well, if after 16 hours of carving and sanding, a spoon snaps. This is after a few weeks of being part of the project. According to Fay, participants began to understand that 'it [was] not their fault.' For most people at the beginning of the project, this would not have been the case.

Whilst sitting outside making with participants, I discovered that for all but one person within this group this had been the first time that they had had experience of creating an object. It was often commented on that it was something special to (a) make an object that could be of use and (b) that could potentially be and would most likely be sold, 'and to someone who wanted it.' When I asked May what it felt like to make things and to have them sold, she responded, 'incredible,' her eyes widening, 'it's amazing.' This brings another complex psychological concept to the fore. The act of making objects in this manner, things that will be sold and valued, created a sense of pride and satisfaction. For all participants, this was a new feeling. A sense of self-worth came with knowing the value of the objects one had made:

> I never usually say this about my own work, they were really good you know and it's the first time in a long time that I can think 'I've made something or I've done something that I'm really proud' of and I'll sit in a group like this and I'll say 'I did that and its good' [...] We just, for whatever reason you set expectations for yourself way above where you should do, but you learn them as a kid or however you learn them you know, its and that made a big difference to me. (Greig)

> I think everyone's felt proud of something that they've done and that grounded-ness as well [...] I feel also not just grounded and not just proud let's say but comfortable and relaxed with feeling those things. [...] I'm constantly feeling excited and feeling good about myself when I'm making something that like turns out quite good and you know and it's not just because other people are complimentary and supportive as well it's because I'm like, 'I actually really like that' and that's OK, like that's, its actually important for me to say that and you know to say I really like the spoon that I've made you know, I'm good at this, like this is great. (Holly)

Silence

Psychoanalyst Wilfred Bion in his discussion of 'the container and contained' speaks of the potentials of empathy and reverie with another.[89] He states that 'at critical moments, to develop thoughts realistically, the human self needs the social container of pair and group to actively participate in meaning making.'[90] For this reason, open conversation and the act of silence within a group had potential in coming to understandings of these group experiences. Often within therapy, a therapist will use 'silence to establish a group culture which values curiosity and verbal communication.'[91]

Whilst making alongside this group it was evident that, in sharing in space with others, there seemed no need to converse. The group was often quiet, occasionally breaking silence with odd comments and tea rounds. One might often think of group development being a vocal and rather physical endeavour. Here however the development of the group was based around sitting, *being*, and making quietly. The group spoke of being happy to sit within one's own thoughts, comfortable with this silence in these environments. Holly explained to me that there was a difference between positively being with one's own thoughts here and elsewhere and why she did not like being with her own thoughts outside of this environment.

Alongside open conversation and observant participation, I chose to utilise the silence that I found within these moments and within some of my excursion activity. Silence had potential in terms of methodology. For Bird-Rose,[92] this might also be described as active listening or 'the work participants engage in as they interact with other sentient creatures.'[93] Bird-Rose, among others, has inspired the proposal of silence as an active verb, meaning to pay attention, to let emerge, and to remain open to experience that may come alongside an allowance for contemplation. This was true for informants who, due to having ample time for reflection, were open in conversation when it did emerge through this silence.

Encounters were often intimate and tactile. They were instances where the participants manipulated and shaped natural materials, whether that be in making or managing and conservation. Whilst some case studies afford the landscape for activity and action and others for more explicitly psychological endeavours, others use their agency to make forms. Through simply *being with* the groups, there was scope for new ways of knowing what these experiences offered to individuals and groups in terms of transformation—providing insight into how these spaces become 'places' of significance endowed with meaning and value.[94]

The case study groups that I have introduced have had a primary interest in the utilisation and instrumentalising of the landscape in order to develop personally. In the case of the young adults, the purpose was to visit unfamiliar places and to utilise the landscape for personal challenge. In the case of The Woodland Weekend Group the purpose was to manipulate the landscape and to reap positive benefits as a by-product of this engagement. For The Community Living and The Mental Health Initiative, the natural landscape was instrumentalised in order to reflect on personal circumstances in a safe environment. I have addressed the ambiguity of these new experiences but from the perspective that we as humans are the initiators of our own experiences within these spaces, whether as an individual or as part of a changing group dynamics.

Becoming a Member of the Community

> I don't know, I think in my response to it, I feel like I've got more empathy with living [non-human] organisms and feel more like maybe this is not completely in line with the Buddhist perspective but maybe, maybe they've got more soul, living things… I think that all sorts of things feel but I don't know, I really can't know about any of this [laughter] (Helen)

My final case study differs somewhat. For some within this group, the non-human and human are far more communicative. The non-human is considered as potentially intentional and an intangible power over experience. The Loose Community are different from previous case study groups because they have specifically come together due to a shared ethos towards natural space and community. The group is made up of multiple practitioners with various related and diversely appropriated skills and interests. Alongside Helen, the psychotherapist and neo-shamanic practitioner and Jill who is a linguist, Finlay is a mental health practitioner and leader of yearly vision quests. Christie is a dancer and arts administrator. Ryan is a community and social justice worker and Goethean scientist. Alice works in a university setting and is familiar with outdoor education and Catherine is a mindfulness practitioner. This is the core 'Loose Community'; however, we were often joined by others who had been brought along or invited. For this reason, it was feasible for me to integrate as a member of the community. The distinction of myself as a researcher quickly faded.

With my other case study groups, I believe that shared intentions came during and after experience. With this group, shared intentions seemed to have come prior to the excursions. A sense of collective obligation felt rather prominent—an obligation to have a meaningful experience, a connection to the landscape, or a psychologically or 'spiritually' important experience.

> parts of our relationship with the natural and maybe also the rest of the world [or] maybe understandings from shamanism […] but it's not the only thing that informs it [her practice]. […] people have been going outside in wild places for spiritual reasons […] I wonder if there was a name for the ancient religions then maybe that would be shamanism? […] I

> wonder if there's been a kind of development towards those nature practices into those religions like Christianity and Buddhism [...] I think that there's quite a lot of shamanic roots in religions and spiritual practices [...] with the kind of neo-shamanism that's taught now [...] I think it's really useful [..] to me, and [...] on an understanding of the world [...] (Helen)

There is often an implicit and sometimes explicit reference towards 'spirituality' within this group. Individuals within The Loose Community are familiar with shared experience within natural environments. They have 'knowledge' of the space and a series of preconceived ideas regarding the benefits, inspiration, and enlightenment that they consider the landscape to offer.

Solo Experiences

At an excursion with The Loose Community in Callander near Stirling, we stayed at Lendrick Lodge—a 'holistic retreat and spiritual centre.' This was a space booked by Helen and Jill to host a nature and creativity weekend. One evening, we lay in our conservatory room with our eyes closed. Helen spoke as Rodica drummed:

> I want you to walk within a natural space, towards a threshold, a gate, two trees, carrying a load and to leave what we don't want on the other side. You will walk through this threshold. You will meet someone; a river, a tree, a wise person, and engage with them [...drumming continues for half an hour]. (Helen (paraphrased))

The intention for the next morning was to repeat this envisioned journey as part of a solo experience in the woodlands of the lodge. In the morning, before our walk we were asked to write a series of words on pieces of card and lay them on the floor. These words were to have come from our 'dream vision journey' the night before. After this, we headed out for five hours to physically recreate our journey alone. When this was complete, we returned to discuss our experiences and to begin creating something that had a connection to our experience. We were invited to partake in this 'dream vision,' where images were painted in our mind's eye. We were encouraged to journey somewhere, where it was hoped that we would feel something, an example of collective intention, or for some an obligation. We then returned to the dry warmth of our conservatory, or temporal normality where we spoke of our experiences. The ethos of shared reflection was useful methodologically.

Creating Objects and Reflection

We had been asked to gather objects from the landscape in order to create something. This was to be reflective of our experience. Throughout this eight-hour making session we were silent, discouraged from conversing. I chose to carve pebbles that I had found in the river where I had spent most of my five-hour solo. I carved symbols representing the various words and images that had appeared in my mind's eye the evening before—a parting river with land in the middle, a heart, a tall woman, a swirling wind, and a shelter. The images within my vision were actively sought out on my walk, and indeed I did find them—the river spot split in two around a small mass of land. The heart, I discovered in the shape of a rock in the river that I tossed back. The tall woman, I spotted by way of a silver birch tree. The swirling wind, in the windswept leaves and the shelter, a curved conifer. This was a strange in-between of obligation and reality. I was told to recreate this journey and to find these images and, I did, to some extent, but to what extent was this a construct of obligation?

In the image below, Finlay has just spent around eight hours slowly binding this dead tree branch in wool. He wound reds and blues in lines around it. He had carried it down the hillside from an area he named 'the tree cemetery' during his solo walk. 'The tree cemetery' was an area of charred and broken trees. Finlay exemplifies a personification of the non-human. He is replanting the dead branch in the ground.

The branch was eventually tossed over into a river and 'allowed' to take its own course. This specific interaction could be considered as ritualistic. Of all case study groups, The Loose Community is the only one to display overtly ritualistic activity. Where some case study groups may show signs of similarities to ritual in their behaviours, such as specific reflection times and explicit structures, repetitive tasks given specific meanings or tokenistic gestures, The Loose Community incorporates these kinds of contemplative and symbolic acts into all of their excursions. A recurring feature is the concept of returning objects to the landscape.

The group considers our interaction with the living and non-living 'more-than-human' as relational. Described by Ryan as 'mutual reciprocity'—a relational dynamics is one in which the group might learn from the non-human. This is a relationship in which the non-human has enchanting qualities. This group chooses to look for and aim to understand new ways to communicate with, and about, what others may deem to be passive or inert objects and spaces in an empathic manner. In some cases, I may say this goes a little further in that informants surrender their own agency over their experience to the 'more-than-human.' The possibility of felt or indeed experienced mutual relationships with the non-human was recognised due to interactions with this group, and a comparison began to form regarding other case study groups. The perception of agency within shared experiences of the natural landscapes then became a key theme for exploring. The realisation that for some the landscape and its non-human material actors have not only an effect on experience but an intention to affect experience or at the very least to be communicative became a major area to explore further.

Throughout my fieldwork, becoming responsive to *being with* my informants in these diverse scenarios became key to my research. For some informants, particularly within The Loose Community, this was not simply the literal being within a space but was in reference to an emotional attuning to one another and the environment. This was an 'awareness' often described as 'sitting with' with reference to dynamic thought processes, arising physical sensations and emotions within these circumstances. I needed to be flexible, spontaneous, and open to contradictions and serendipitous moments that may allow for further enquiry, probing and understanding. If I had approached my fieldwork in any other manner, the themes that were able to emerge may not have, they may have been stifled by overpreparation and structure. They may have been lost in an uncomfortable and unequal informant-researcher relationship.

In socialising with my groups, in immersing myself in participation, in allowing myself to be guided by the frameworks and structures already present within these groups and their activity, I was able to gain an insight into the way that the group dynamics shifted. I was immersed in the way that people interacted and related to the group and their environments and was considered temporally a part of the group. I was involved in reflection, downtime and what may otherwise have been private conversations. Within fieldwork I was shown around. To get to the locations, I journeyed as informants did, and I did as informants did by fully participating in all activities. I integrated myself, shared spaces, shared food, and shared sleeping areas, and I built upon relationships before attempting one on one conversations—this proved fruitful. By taking on a role other than researcher, I gained perspectives from participants, practitioners, staff, and volunteers. I was open to contradictions and to exploring those contradictions. I walked with informants and was idle and tactile. I volunteered with informants, taking part in their context, and I tried to learn. I was taught how to do things. I was silent and utilised this silence. I reflected with the group, and I became a part of these communities—it was through these *methods* that I gathered my data and allowed themes to emerge.

Within this chapter, I have provided a more realistic account of research methods used within the field. They were not predetermined but instead emerged through interaction, full participation, and being present and responsive. The methods that emerged were serendipitous. The opportunities for gathering data presented themselves, and I utilised them as methods. Each method was dependent on the case study group context, the activities that they engaged in and the individuals within the groups. These methods were successful for me because they were successful within context.

Notes

1. These methods were adopted from The Natural Change Project by the facilitator. I felt that my story was no longer my own in this instance—a fate I would not wish to impose on informants.

2. 'Flow' (Csikszentmihalyi 2002) describes the phenomenon of 'being in the zone,' of times of total immersion in the task at hand when an individual's body and mind are stretched in an effort to accomplish something worthwhile: perhaps flow may have been encountered within these experiences. In flow, one is said to be in a space where 'time stops,' where meaning and purpose is out with the self (Seligman 2011). An integral aspect of flow as in peak experience is a belief that by investing all psychic energy into interaction, one 'becomes part of a system of action greater than that which the individual self had been before.' (Ibid: 65).
3. Kahneman and Riis (2005).
4. Kahneman and Riis (2005: 285).
5. Kahneman and Riis (2005).
6. Kahneman and Riis (2005: 286).
7. Kahneman and Riis (2005: 286).
8. Kahneman and Riis (2005: 287).
9. (Moeran 2007: 04).
10. Moeran (2007: 12).
11. Moeran (2007).
12. Moeran (2007).
13. Moeran (2007: 20).
14. Moeran (2007: 13).
15. Moeran (2007: 15).
16. Wilkinson (2017).
17. Wilkinson (2017: 619).
18. Wilkinson (2017: 614).
19. Wilkinson (2017: 619).
20. Ingold (2018: 59).
21. Ingold (2018).
22. Ingold (2018: 61).
23. For a more step-by-step outline of what may be considered as walking and talking as a method, see the 'Go Along interview' (See Kusenbach 2003; Carpiano 2009) more simply described as 'on the hoof' conversation by Pinder (2007: 116).
24. This is similar to hanging out or 'deep hanging out' (Geertz 2008)—spending downtime or unstructured time with informants, chatting, observing, and so on.
25. Madden (2013: 01).

26. see Kusenbach (2003) and Carpiano (2009).
27. Rivioal and Salazar (2013: 180).
28. Merton 1948 in Rivioal and Salazar (2013: 178).
29. See Glaser and Strauss (1967).
30. Hannerz (2006) in Rivioal and Salazar (2013).
31. in Rivioal and Salazar (2013: 183).
32. Madden (2013: 16–17).
33. Rivioal and Salazar (2013).
34. Fetterman (2010).
35. Rivioal and Salazar (2013: 02).
36. Rivioal and Salazar (2013: 178).
37. Rivioal and Salazar (2013).
38. Law (2004).
39. Le Courant (2013: 186).
40. Le Courant (2013).
41. See Russell et al. (2007).
42. Ingold (2018).
43. Le Courant (2013: 195).
44. Le Courant (2013: 195).
45. Le Courant (2013).
46. Davis (2007).
47. Davis (2007: 09).
48. Davis (2007: 11).
49. Davis (2007: 02).
50. Davis (2007: 02).
51. Davis (2007).
52. in Hazan and Hertzog (2012).
53. Ben-Ari (Ibid: 67).
54. Davis (2007).
55. Herzfeld (2012).
56. Herzfeld (2012: 114).
57. Herzfeld (2012: 114).
58. Hazan and Hertzog (2012).
59. Fabietti also speaks of an interdisciplinary career trajectory and another, Shapira also discusses anthropological serendipity in relation to her career (both in Hazan and Hertzog 2012).
60. Goethe (1823/2010: 19).
61. Greverus (2002: 1).

62. Greverus (2002).
63. Greverus (Ibid: 09).
64. Goethe published his first scientific work in 1790. Though previously he had been considered a philosopher, dramatist, poet, and statesman, he penned *The Metamorphosis of Plants* (1790/2009). This work is associated with Goethean observation, a way of observing and understanding plants and the non-human natural world.
65. Goethe (1823/2010: 19).
66. Greverus (2002).
67. Greverus (2002: 41).
68. Greverus (2002: 10).
69. Goethe (1823/2010: 20).
70. Ingold (2010: 135).
71. Lewis (2000: 54).
72. Ingold (2010).
73. Senda-Cook (2012).
74. Senda-Cook (2012).
75. Senda-Cook (2012: 144).
76. Philo (2003).
77. Philo (2003: 12).
78. Rousseau (1783/2011).
79. Rousseau (1783/2011).
80. Philo (2003: 07).
81. Bachelard (1960/1969).
82. Berleant (1992: 14–15).
83. Berleant (1992).
84. Berleant (1992: 14–15).
85. Bloch (2012).
86. Bloch (2012).
87. A Scottish stirring utensil.
88. A tool for making holes in the ground to place seeds into.
89. Bion (1970).
90. Bion (1970: 246).
91. Bion (1970: 248).
92. Bird-Rose (2013).
93. Bird-Rose (2013: 93).
94. Tuan (1979).

References

Bachelard, G. (1969/1984). *The Poetics of Reverie – Childhood, Language and the Cosmos*. Boston: Beacon Press Books.

Berleant, A. (1992). *The Aesthetics of Environment*. Philadelphia: Temple University Press.

Bion, W. (1970). *Attention and Interpretation*. London: Tavistock.

Bird-Rose, D. (2013). Val Plumwood's Philosophical Animism: Attentive Interactions in the Sentient World. *Environmental Humanities, 3*, 93–109.

Bloch, M. (2012). *Anthropology and the Cognitive Challenge*. Cambridge: Cambridge University Press.

Carpiano, R. M. (2009). Come Take a Walk with Me: The "Go-Along" Interview as a Novel Method for Studying the Implications of Place and Health in Wellbeing. *Health and Place, 15*, 263–272. https://doi.org/10.1016/j.healthplace.2008.05.003.

Csikszentmihalyi, M. (2002). *Flow: The Psychology of Optimal Experience*. London: Rider/Random House.

Davis, D. (2007, October). *Science and Serendipity: Reflections on an Anthropological Career*. 55th Annual Harrington Lecture (pp. 1–27). Vermillion: The University of South Dakota.

Fetterman, D. M. (2010). *Ethnography: Step-by-Step*. Los Angeles: Sage.

Geertz, C. (2008). Deep Hanging Out. *The New York Review of Books, 45*(16), 69–72.

Glaser, B., & Strauss, A. (1967). *The Discovery of Grounded Theory*. Hawthorne: Aldine Publishing Company.

Greverus, I. M. (2002). Anthropological Voyage of Serendipity and Deep Clues. *Anthropological Journal on European Cultures, 11*, 9–50. Shifting Grounds: Experiments in Doing Ethnography. http://www.jstor.org/stable/43234893

Hazan and Hertzog. (2012). *Serendipity in Anthropological Research: The Nomadic Turn*. Oxon/New York: Routledge.

Herzfeld, M. (2012). Chapter Five: Passionate Serendipity: From the Acropolis to the Golden Mount. In A. Gottlieb (Ed.), *The Restless Anthropologist: New Field Sites, New Visions* (pp. 100–122). Chicago: The University of Chicago Press.

Ingold, T. (2010). Footprints Through the Weather-World: Walking, Breathing, Knowing. *Journal of the Royal Anthropological Institute, 16*, s121–s139.

Ingold, T. (2018). *Anthropology and/as Education*. Oxon: Routledge.

Kahneman, D., & Riis, J. (2005). Living and Thinking About It: Two Perspectives on Life. In N. Baylis, F. A. Huppert, & B. Keverne (Eds.), *The Science of Well-Being* (pp. 285–301). Oxford: Oxford University Press.

Kusenbach, M. (2003, January). *Street-Phenomenology: The Go-Along as Ethnographic Research Tool.* PhD Thesis, University of California, Los Angeles, Department of Sociology.

Law, J. (2004). *After Method: Mess in Social Science Research.* Oxon/New York: Routledge.

Le Courant, S. (2013). What Can We Learn from a 'Liar' and a 'Madman'? Serendipity and Double Commitment During Fieldwork. Young Scholars Forum. *Social Anthropology, 21*(2), 186–198. https://doi.org/10.1111/1469-8676.12015.

Lewis, N. (2000). The Climbing Body, Nature and the Experience of Modernity. *Body and Society, 6*(3–4), 58–80.

Madden, R. (2013). *Being Ethnographic: A Guide to the Theory and Practice of Ethnography.* London: Sage.

Moeran, B. (2007). *From Participant Observation to Observant Participation: Anthropology, Fieldwork and Organisational Ethnography.* Creative Encounters Working Papers. No. 2 (pp. 1–25). Copenhagen Business School.

Philo, C. (2003). 'To Go Back Up the Side Hill': Memory, Imaginations and Reveries of Childhood. *Children's Geographies, 1*(1), 7–23. Carlax Publishing Group. https://doi.org/10.1080/14733280220000041634.

Pinder, R. (2007). On Movement and Stillness. *Ethnography, 8*(1), 99–116. https://doi.org/10.1177/1466138107076142.

Rivioal, I., & Salazar, N. B. (2013). Contemporary Ethnographic Practice and the Value of Serendipity. Young Scholars Forum. *Social Anthropology/Anthropologie Sociale, 1*(2), 178–185.

Rousseau, J. J. (1783/2011). *Reveries of a Solitary Walker: A New Translation by Russell Goulborne.* Oxford: Oxford World Classics, Oxford University Press.

Russell, A. W., Wickson, F., & Carew, A. L. (2007). Transdisciplinarity: Context, Contradiction and Capacity. *Futures, 40*, 460–472. https://doi.org/10.1016/j.futures.2007.10.005, 2008.

Seligman, M. (2011). *Flourish: A New Understanding of Happiness and Wellbeing – And How to Achieve Them.* London: Nicholas Brealey Publishing.

Senda-Cook, S. (2012). Rugged Practices: Embodying Authenticity in Outdoor Recreation. *Quarterly Journal of Speech, 98*(2), 129–152. https://doi.org/10.1080/00335630.2012.663500.

Tuan, Y. F. (1979). *Space and Place*. London: Edward Arnold Publishers Ltd.
Tuckman, B. W. (1965). Developmental Sequence in Small Groups. *Psychological Bulletin, 63*(6), 384–399.
Turner, V. (1969). *The Ritual Process – Structure and Anti-Structure*. Suffolk: Richard Clay (The Chaucer Press Ltd).
Turner, V. (1970). *The Forest of Symbols: Aspects of Ndembu Ritual*. Ithica: Cornell University Press.
Turner, V. (1974). *Dramas, Fields, and Metaphors*. Ithaca: Cornell University Press.
von Goethe, J. W. (1790/2009). *The Metamorphosis of Plants* (trans: Miller, G. L.). Cambridge: MIT Press.
von Goethe, J. W. (2010/1823). The Experiment as Mediator of Object and Subject. *In Context 24*, 19–23, The Nature Institute.
Wilkinson, C. (2017). Going 'Backstage': Observant Participation in Research with Young People. *Children's Geographies, 15*(5), 614–620. https://doi.org/10.1080/14733285.2017.1290924.

4

The Journey, Belonging, and the Self

I trudged up Waterloo Place, a long, steep extension of a bustling Princess Street in Edinburgh, lined with shops, cafés, taxis, hotels, and past the train station. It was November. I had my haphazardly packed bag of essentials, gathered from a kit list provided, and I was heading to our pick-up point, the roundabout at the old Royal High School. I saw Claire surrounded by bags, some plastic carriers, and some hardy, her hands in pockets, slumped against the wall. She was wearing a baggy black fleece and torn waterproof trousers. She glanced at me shiftily—an awkward standoff on the street corner. I wasn't sure if she was waiting for what I was waiting for. We stood below Edinburgh's Calton Hill and overlooking Arthur's Seat, perhaps a hint towards the kind of landscapes we might have been journeying to, only this time, beyond the urban streets of where we were now. We were going to be collected by a minibus, I supposed, though I wasn't entirely sure. I wasn't certain where we were going or how far we might have to travel.

Navid and Aasim strode up the hill towards us, hiking, backpacks held tight by the straps, sleeping bags slung over their shoulders. Navid is 25, a 'British Indian' who grew up in London. He has been coming to these excursions every month for a year and a half. He has just resigned from the family business, producers of wholesale foods, and has no idea where

R. Crowther, *Wellbeing and Self-Transformation in Natural Landscapes*,
https://doi.org/10.1007/978-3-319-97673-0_4

he will go next, or so he told Aasim, and then again Rosa who had just joined us. They chastised him: 'What happened?! But it's family! What did your dad say?' Aasim is a PhD student from Libya; he studies engineering, and Rosa, whose mother is from Mozambique and whose father is Northern Irish, considers herself as representative of both cultures. She has lived in the United Kingdom for 20 years and works in human resources at the university.

Rosa had been in a long-term relationship, and when it ended, having 'neglected her friendships,' she found herself with nothing to do. She began coming along to these excursions because she 'had no friends' and wanted to 'try new things.' They greeted Claire. They had clearly met many times before as they chatted animatedly about previous excursions and the people who hadn't arrived yet. Claire has been coming to these excursions for years. In her late 30s, she has learning difficulties and is considered by the group to be a key member. She welcomed everyone and reminded them of what they last spoke to her about. Claire grew up in East Lothian. She was keen to keep an eye on the football score and made sure that she found ways throughout the weekends to let George know too.

George meandered into the group, an older man of around 70. He had already been for a long walk around Edinburgh, as he did every time. He had walked up from that station every month for nine years to meet these groups. George joined the group from Newcastle. He always did. He planned his trip months in advance—three pounds direct with his pensioners' travel card. He has volunteered ever since he retired from the telecom engineering business.

The group began to grow larger. I had met Carolyn and Simon previously at a wood fuel collection day, when they had dropped in to join the group for lunch in the woods. They are 'regulars,' something that causes a little contention amongst the group due to 'over-regulars' sometimes making 'newbies' feel unwelcome. As Navid said, 'We were all new once, they need to remember that.'

> How do new people fit into it? not always as well as I would like them to, to be honest, I think when you have a lot of people who have been coming for a long time… I mean we go through phases, we're in a moderate phase

> but we had a phase, you know phases where there's lots of them [regulars.] I think it's off putting for new people and I think they only come once and they maybe don't come back, you know? and they haven't had a terrible time but it hasn't risen above for them to have a great time and so I think that's quite challenging because I think over-regular people, for want of a better word become quite set in their ways and perhaps the fact that they are so regular is because they're already set in their ways and they, they can have a kind of controlling… a desire to control the way things are and that can cause conflict and that does happen with the woodland weekends sometimes. (Ross)

There is a certain level of initiation within this group—my feeling was that it was somewhere between the third and fourth excursion, and this is down to the control that some within the group would like to have. 'Their thing' is guarded quite closely, and suspicion of 'newbies' is common. It took some people within the group some time before beginning conversations with me. It took Claire, Carolyn, and Simon several excursions before they began to speak to me. The evenings spent socialising over whisky and wine sometimes helped with this. Carolyn and Simon have volunteered for several well-known Scottish conservation charities. Their *better* knowledge and experience is often flaunted and is sometimes expressed in an overzealous felling of trees.

Ross arrived around 6:15; he is notoriously late. We piled our backpacks, sleeping bags, and mats in amongst the giant Tupperware boxes of food—bread, milk, tea, lettuce, cheese—into the back of the minibus. It was tight for space. I clocked a 'newbie.' She stood to one side, seemingly keen to chat but holding back. She told me later, when we met in the pub, that she struggles in group situations and turns red as soon as she speaks. I empathise with her. It's an odd thing to meet a group of strangers and prepare to jump on a bus with them to a mysterious location:

> I was the first on the bus and went straight to the back. I thought this might have been a mistake as I couldn't talk to people. It was dark as well which is weird. I thought, I should be chatting more, but I can't. Some people fell asleep. It is hard, I was apprehensive. (Isla)

People chatted on the bus and lulled in and out of silence—a few fell in and out of sleep. Isla and I sat at the back of the bus attempting to eat fish and chips in the dark using the light from our mobile phones. The phones would soon be made redundant when we arrived. With all case study groups, mobile phones were either not allowed or unable to get phone reception. Often the phone became only a means for a few snaps before it ran out of battery, and there was no way of charging them.

We travelled for around three hours to a village hall on the outskirts of Grantown-on-Spey. When we arrived, everyone hopped out and dove straight into the hall. They seemed to know exactly what they were doing. I loitered at the back of the van as Ross offloaded bags. People came back and forth to grab things. I got the impression they were more concerned with nabbing the best sleeping spot on the hall floor. I realised this when I finally got inside and found that the only available space was near the fire exit door that would be kept open all night, with strip lighting above, for those who may struggle to find the toilet in the dark. I stood watching, a little unsure of where to start and anxious that I probably wouldn't sleep that night. Others were busy moving chairs, improvising makeshift barriers, and blowing up mattresses.

> I imagined a wee bunk house, a wee hut up a mountain and a bit more basic. But instead we were on the floor [...] It's somebody you've just met and now you're sleeping next to them. [...] It was funny. People have their own ways. When you're alone you don't realise that you have your own quirks. You don't know people and you're already seeing quirks, even their sleeping habits. I was just in the middle [of the room] I just didn't care and was like 'night guys!' its odd waking up to a stranger. (Isla)

For Isla, the sleeping arrangements were novel, but as a self-proclaimed backpacker, this novelty was welcome. In fact, Isla had turned up to this trip with very little, a sleeping bag only, not quite up to the hardwood floor. I gave her my sleeping mat. For Rosa, the sleeping arrangements were a stressful anticipation. On every excursion on which she may be required to share a sleeping space, where, how, and with whom she was sleeping next to was discussed anxiously until special arrangements were made for her. This time Rosa chose the stage area with Robert, who we had collected on our way, from Perth station.

Robert is the eldest within the group and is considered a little like a friendly uncle. He referred to Navid and Aasim as his sons and regularly encouraged them to 'escape off' to the pub with him. Rosa had a real difficulty sleeping if she could 'hear others breathing,' and as Robert doesn't snore, he was 'allowed' to sleep up on the stage, along a row of cushioned chairs. Rosa had slept on this stage before and had made a beeline straight for it. She went as far as to tell us that she may have had to sleep in the small cupboard under the piano that night. It was barely big enough for a camping mat. We all slept in this village hall—creaks, sighs, wriggling, strip lighting, and snoring all night long. In the morning, we began work on clearing a community woodland of non-indigenous trees (Fig. 4.1).

Fig. 4.1 My sleeping area in the town hall

Sleeping arrangements on these excursions were often very improvised, and often people changed their sleeping space throughout the weekend to obscure spaces that they found elsewhere on the site. One bad night's sleep forced people to look for somewhere else, anywhere else. Even if that somewhere was a vegetable planter or, for me, a set of kitchen chairs. The aim of the game was to sleep, even if only a little in such unfamiliar surroundings.

Back to Basics in New Spaces

Aspects of these excursions including the 'different from the norm' sleeping situations, simple foods, lack of access to the internet, and redundancy of technology were key elements of these experiences for informants. This 'back to basics' concept is evident in all of the groups. For some, this is what they seek; for others, this is the most challenging aspect (Fig. 4.2).

Fig. 4.2 Navid's bed in the poly tunnel

> I think there's a real mystical quality to these trips and I'll tell you why that is, you meet in Edinburgh, outside the bus and you congregate outside the bus and, you know the name of the place you are going and you've probably done some research or google imaged it before but for the most part you've probably never been before or you …it's the kind of adventure aspect of it, you're going to this place, you don't know what the sleeps gonna be like, you don't know the kind of work until you're gonna be there, that element of surprise I think is attractive. (Navid)

The mystical quality described by Navid is something that was mentioned to me by several informants that travel on excursions with this group. They described the feeling of not knowing, the expectation, and the new additions to the repertoire of lands that they had worked on previously. The minibus was once described to me as the 'magic bus, you just jump in and it takes you somewhere, all you need to do is get on' (Tom). This man revelled in the idea of jumping on a bus and ending up in 'the middle of nowhere.' It was the sense of 'adventure' that came with leaving familiar urban surroundings and heading somewhere new with people who were not a part of your day-to-day life that enticed the majority of this group. Part physical, part emotional, and part social, it seemed that these people wanted to experience something new and in doing so be able to bring *something* back to their daily lives in the city.

The Journey and Role Development

When travelling with The Young Adults' Personal Development Project, the minibus journey is perhaps the most contentious part of the excursion. This was often where relationships and roles were established. The young adults were often a fragile balance between apathetic and hyper. They had always planned the excursion in advance, having discussed where they might be going or what may be happening with the facilitators. Often the locations and activities were altered at the last moment, adding to a sense of disarray and to the complex group dynamics.

In the minibus people sat very close to one another. It was dirty, damp, rickety, and with a maximum speed of 50 miles an hour. The radio was always on, chart songs on repeat, blasted in the back through the tiny speakers above our heads. Some chose to sit alone on single seats, others chose to share or were asked to share a space. Whilst some sat quietly, others shouted profanities out of the windows or over one another to the volunteer support workers in the front seats. The journey was a key stage in the development of the group. It was often tense, and anxiety ridden, sometimes jovial, and often a space of challenge and confrontation.

This journey was the beginning of these excursions away from the urban. It is a necessity of travelling in groups and a space that could be considered as somewhere in which group dynamics change and where group roles emerge. This happens whilst we move through the landscape and towards a space out with informant's comfort zones. This nod towards liminality is obvious. We were travelling from a place of familiarity to one unfamiliar, and then returning again, along with the ambiguity that comes with this. This is an important aspect of the excursions. For all activities and across all case study groups there was a journey—a spatial journey from within an urban environment to a natural one.

In order to reach these landscapes, it was sometimes required to travel by other means. This was the case with The Community Living Initiative in the Inner Hebrides. Groups and individuals needed to travel across not only land but the sea as well, a liminal physical landscape:

> Being on an island you have to journey to get here, you have to get here by ferry and you have to go [big intake of breath] by bus and then you have to walk down the track to get here so there is a real sense of movement and destination and effort to get here which I think enhances the thing that it's not just somewhere you roll up to on the bus and get out and go straight to your dorm and things, you actually, it takes time to get here and during that time people have experiences and they've got something to look forward to and… it's almost like it gets more isolated, more remote as you come, you know you, come on a train, people have been on a train before, and then they get on a ferry, they may not have been on a ferry before, then going on a bus through the mountains in through the glen and that's a different experience and then they get [here] and then they've still got to walk a mile and a half [laughter]. (Allan)

It is clear within this instance that the journey itself was a significant part of the experience. The journey takes over 10 hours, travelling by train, ferry, bus, and foot to the site of the initiative. Some individuals that visited were coming from cities within the central belt of Scotland. They were witnessing an essence of what Scotland as 'wild' may have been like for the first time. For many, it was their first time without internet connection, electricity, and regularly timed public transport. There seemed to me to be far more to this than travelling across physical landscape boundaries. Within these journeys, there were personal challenges, new experiences, unfamiliar territories, and new cultures. Some informants were already 'part of this community'; however, some were coming from economically deprived locations, as part of social organisations. Some had been here many times before, and others were discovering it for the first time (Fig. 4.3).

Fig. 4.3 New landscapes to experience at The Community Living Initiative

When group facilitators explicitly stated that these journeys were for personal development, the participants tended to come from a lower socio-economic background. Participants both in The Young Adults' Personal Developmental Project and The Community Living Initiative, prior to engagement with these projects had had little interaction with natural space. Their agendas for doing so were often to do with furthering employability, overcoming personal challenges, and pursuing skill development. This too had a significant amount to do with taking urbanites out of their comfort zones and into landscapes of which they were likely to be unfamiliar.

The Journey Metaphor

Informants speak of physical, psychological, and spiritual journeys, crossing personal boundaries, challenging oneself, taking oneself out of one's comfort zone, and becoming part of new groups. The informants that I have discussed above seemed to be not only between spaces physically but between metaphorical 'spaces' or stages in their lives and in their ways of thinking—between jobs, growing up, looking for 'something,' seeking relationships, and wanting to try 'something new.' These spatial and metaphorical journeys, when entwined, were considered as intangible and ineffable. This spans the interactions with the space and with others, as well as the effect that this has on the self.

> [...] the belongingness hypothesis is that human beings have a pervasive drive to form and maintain at least a minimum quantity of lasting, positive, and significant interpersonal relationships.[1]

If one wishes to speak of transformation, then knowing what it was that attracted individuals to the projects of which they were part is important. Each person's motivation, agenda, wants, and needs affected the dynamics of the group whether they intended to become more employable or whether they intended to learn new skills or to build new social relationships. Whilst each person is unique, and each has their own subjective reasoning, people are driven in part by the desire for a sense of belonging.

People were motivated by belonging in that they wanted to be part of a group. Individuals wanted to 'feel in tune with one another,' 'have people to talk to,' to be part of a social network, and to find a place to be where one 'would be missed if they were not there.' Later, I will link this with a desire to fulfil an aspect of one's self-identity. The motivation to belong and to self-identify with the group, activity, and landscape preceded observed transformation and subsequent feelings of mental wellbeing. These ultimate motivations affected individual relationships with the group, the group dynamics, and the perception of activities and the space. It was due to this relational dynamics between a desire to belong, to finding that sense of belonging, and being within these environments that these excursions provide a distinctive site that fosters satisfactory opportunities for belongingness.

A journey may not be intrinsically transformational. Some experience self-verification and others don't. Some experience a sense of 'communitas,' or at least some of the aspects that Turner[2] discusses when he speaks of 'communitas,' and others may not. However, I believe that the motivation to engage in these activities was still a desire to belong within a group and ultimately to verify an aspect of one's ideal or ought sense of self. An individual may have multiple intertwined motivations to take part; however, throughout my time with groups several motivational themes emerged that were separate to the agendas expressly put forward by the groups' frameworks and ethos. By looking at the individual, joining and becoming a part of a group of people and at how the collective of individuals developed throughout the journey, activities, and time spent together within these spaces, I discuss how the groups manifest a bonded and cohesive dynamics. This dynamics characterises aspects of Turner's 'communitas.' The desire for a sense of belonging drove groups towards this sense of 'communitas.'

New Cultural Interactions

For Rosa, it is rare that as adults we are able to involve ourselves in activities that provide truly impartial opportunities to 'do something different' and 'to meet strangers.' For George, who wasn't only getting out of town but was crossing the country to attend excursions in Scotland, it was the remoteness of some of the locations that was the 'main attraction.' It was a 'break from normal life.' It was something 'completely different.' This subversion of the norm was found in not only actively seeking something different—leaving one's everyday circumstances or getting out of town—but also in challenging oneself to find a new job, or learn new skills, and so on. This could arguably be the case without the need to be within natural spaces. This could also be carried out as an individual without the need for a group. All of the individuals, regardless of explicit motivation, however chose the outdoors and the social as their framework. This seemed to be about more than subverting the norm. It was about participating in activities or communicating with people that one might deem as reflective of an aspect of themselves (Fig. 4.4).

Fig. 4.4 Rosa in Grantown-on-Spey

> Identity is a complex and contested term. Here, identity is taken to be a performed status, having to be created and maintained in the flux of individual and collective everyday life.[3]

The 'belongingness hypothesis' suggests that people join groups to feel that they belong within a group. The attraction to a group is often based, rather than simply wanting to find a group, any group, on finding a group that may correspond with one's own ideals, interests, morals, spiritual identity, or political standpoint, and so on. Whilst we might argue that underlying all motivation to take part in group activities in natural environments may be the innate human desire to belong, individual motivations emerged within the field as complex. This stems from this innate human desire or perhaps obligation to be part of normative societal structures, to be part of a group of like-minded people, or to be part of a perceived community.

> [the] job centre sent me. (Paul, Sam, and Amy)

For the young adults joining the personal development project, a primary goal, as set up by the framework of the organisation, was to become more employable. In this specific instance, we might say that Paul, Sam, and Amy would like to get themselves a job on completion of their time with the charity. Presumably this would be to adhere to societal norms and to fit in with the status quo. However, because all three of these young adults on separate occasions told me that they would like to be a permanent volunteer within *this* organisation and that they felt comfortable within *this* group of people, one could argue that this may not be entirely the case. The initial obligation (for several their response to my asking what brought them to the group was that the Job Centre had sent them) may have been felt as, *I need a job*. It was clear to me, however, that for these young adults the true commitment to the project lay somewhere else. All the participants within this project were, at my time of meeting them, unemployed. The young adults had, for the majority, little work experience and few, if any, had formal qualifications. Sam does not currently work, and he says;

> [… it's] something to do—get out the house […] do I go or do I just sit about the house all day? […] even though you can't be bothered and you don't want to do it, you still [come along]. (Sam)

A driving force behind Sam's attendance was: 'Getting out, [and] seeing a bit more of the country.' There seemed to be a feeling of apathy towards the excursions and the project, which I believe may have been a detrimental effect of the differing intentions of both the group and the support workers on his experience of these places. He considered the project as simply 'something to do' but gave the impression that having something, or *anything* to do was a better option than sitting and doing nothing at home. He stated that he often 'couldn't be bothered,' nor did he want to attend but nonetheless he returned. It seemed odd to me that Sam, and others with similar attitudes, seemingly did not want to be at the project; yet they came along each day. A question I often asked, considering the often hostile and argumentative environment was, *what makes them keep coming back?*

Whilst getting to know Sam, I was made aware that every day Amy would come to his house and 'drag him out.' She did her best to make sure he attended, but not because she particularly enjoyed it either, or so she said. Instead, because she would not have come herself if Sam was not there. This reinforcing the undercurrent, even from those who state seeking employment as a motivation, of developing social relationships as a driving factor for taking part:

R: Do you think you would have made it this far had Sam not been here?
A: No, I don't think so like, I would have left after that first five-day trip.
R: Ok fair enough, what about you? […] if Amy wasn't here?
S: I don't know, probably not like, she's forced me to come a few times. (Amy and Sam)

Amy and Sam both refer to Amy having forced Sam to attend. Perhaps 'force' may be too strong a word but nonetheless, Amy wanted Sam to be there, and Sam obliged. There must have been a reason for this, and I believe that contradictory to their apathy, they gained something from being here. Amy had also been advised to go along by her social worker. She is a single young mother and is also unemployed. It was clear to me that Amy looked out for Sam, and I couldn't help but think that she wanted him to succeed. Knowing that Sam was coaxed often by Amy suggests that he felt an obligation to her. Similarly, knowing that she would not have attended without him suggests something of an in-group mentality—a wanting to be with one another but perhaps not feeling that they identified with the wider group. Disdain with others in the group was often vocalised. This was to do with the age and maturity of others, living circumstances, and abilities. It was clear that both Sam and Amy considered themselves as different to the rest of the group:

R: What is different? Why are you different to the rest of the group?
S: I don't know, they still live with their mum and that, like I've had my own house since I was 16 years old, they're just like bairns […] and they start going that immature way and it's just like, 'f**k off man!'
R: Do you think that they are going to get anything different out of this that you might not have?
A: maybe a pair of b**ls? [Laughter]
[…]
R: When you first walked into the room with that group of people, how did you feel and how did you interact with them? do you remember feeling any certain ways?
A: like, what the f**k am I doing here with these people!? [laughter]
S: [I felt] about exactly the same as her
A: Well I don't know, cos it just depends on what the type of person is. Like Fraser and Jonathan, obviously they have autism and they act like, well it makes you like act really young, stuff like that, I don't know…
R: Do you think there's also something in you having a different responsibility that nobody else has? [Amy has a two-year-old son]
A: I think so, I like to be immature and that, well I don't like being immature, I like to, oh I don't know…

Sam and Amy do not seem to identify with the group; however, they have found a sense of belonging and self-verification in each other, with similar beliefs about the rest of the group, and similar life aims. They have identified differences amongst the group. They consider them to be more immature, to have less adult responsibility, and to have learning difficulties. Amy and Sam believe that they are here to get a job, that they are more mature, and that they have more responsibilities. Amy did not believe that she belonged with these people; however, she 'enjoyed [Sam's] company.'

> I think mainly the reason for coming anyway was for having something to do. I mean I was in the house 24/7 and all I would do is play on my Xbox and all that and it's not really healthy. So, I thought if I push myself to the limits and then I can find something that I'm good at. (Jonathan)

Similar to Sam, Jonathan told me that his primary reason for attending the personal development project was to have somewhere to go, only this time specifically to enable him to do more than play his Xbox. He wanted to find something else that he was good at. He acknowledged that staying in his home all day and every day was 'not healthy,' presumably referring to his mental health and the lack of social and outdoors interaction. He suggested that attending the personal development project, in opposition to his everyday scenario, may be a healthy thing to do—the comparison pointing to social interaction and a challenge to do something new.

For Jonathan, this is a public outing that is outside of his norm and allows interactions that he may previously have been inexperienced in. Jonathan associated getting himself out of the house and attending with pushing himself to the limits, regardless of activity, to discover something that he is good at. For Rory, the group facilitator, it is common for participants to attend whose comfort zone is 'literally sitting in a bedroom and playing their Xbox and doing very little else,' a comfort zone exactly as Jonathan speaks of. For Rory though, there is more to this than just wanting to get out of the house. Participants are aware that their situations, whilst perhaps comfortable, are not conducive to a positive future:

> [...] now they wouldn't be here in the first place if they were happy with that. So, that's, you know, that's the first point- It's not that we went into

> their room and actually [said] 'this is bad! You need to do something about this'. They came here and said, 'Can I try something?' (Rory)

Amongst Jonathan's comments, there is a distinct desire to meet others, particularly when he refers to his behaviours as unhealthy. He has no face-to-face social interaction in his life. Previously I spoke of Jonathan's online gaming habits that allowed him to converse with people all over the world and allowed him to feel comfortable in the knowledge that these people did not know what he looked like or that he had autism. He explicitly stated that he wanted to do more than play on his Xbox—to do more than engage with these online communities. He explicitly stated that he was concerned about introducing himself to groups of people who had the potential to be judgemental. Yet, Jonathan wanted to get out of the house because he wanted something different for himself.

Aasim, from The Woodland Weekend Group, explicitly stated that he too strived for a different experience: 'To learn something new, to meet new people.' He drew a comparison between his laboratory and engineering work. He confessed to having no experience with nature, and to how he had wanted to learn and 'to meet people.' In his own leisure time, he cycles avidly. He is also a Scout leader. In advance of the excursions he felt that there may be some synergy between the knowledge that he has already and things that he knows he likes and this different experience.

> I had just come out of a long-term relationship at the time and had found myself quite isolated because a lot of my friendships had kind of fizzled away in the time that I was involved so I found, I was in a place where I wanted to meet new people [...] I go because I like that social interaction you know I sit in an office by myself all day during the week [...]. (Rosa)

Rosa had wanted something completely new: 'I remember just going through the community page on Gumtree and looking for events or sports clubs [...].' She did not specify exactly what she may have been aiming to find on Gumtree; however, she knew what she wanted to find within the activity she was going to take part in. She too works in an enclosed office space and was looking for *something*. This concept of looking for *something* is recurring in conversation within my case study groups—looking for *something* more, looking for *something* fulfilling, looking for *something* to add value to one's life, looking for *something* to provide… *something*. There is a sense that people within these case study groups are seeking more than what they believe they have already; they are looking to belong somewhere, in an activity or with a group. What it is that they seek seems, often, ineffable.

For George, being outdoors is important but it was the social aspect that was key. He told me that this is the reason '[the group] always stay in one place […] always together. [They] work together all day and eat together […] sleep in the same room […] socialise in the evening.' For George, this is the whole point of the experience. The woodland weekends are about being together, and this is valued. For Greig similarly, he enjoyed the laughter and is confident that even on a bad day the group was a support. This was important to him. Holly was more straightforward: 'The group is the thing that makes me come […]' This desire to be social either with others within the group or indeed with the non-human proved to be a key factor in joining these groups.

> […] you can go out in public and be invisible, you can go out in public and be absolutely isolated and feel like crap and nobody will know it's […] that's so sad. That's like the loneliest thing ever isn't it? […] sometimes I notice when I'm here and there is a quiet moment and I am having that headspace, what I've noticed recently is that sometimes this is actually the space where […] I'm able to have the headspace where I'm kind of in a way… […] certain problems that I'm having or certain things that are coming up are actually being processed in this safe space because I'm not alone but I'm alone enough, you know to safely like ask myself some questions or go through some things in my mind and make decisions about what should I do about that thing or how do I really feel about… you know? […] where I'm able to come up with clearer answers for myself whereas if I'm alone, alone and I don't know that someone's gonna bring me out of it or you know, or miss me if I'm not there, then I will tend to go into escapist mode and I will not address any of my things because it just feels too overwhelming […] (Holly)

Holly expresses the difficulties in feeling alone and 'invisible' when suffering from depression and anxiety within social groups and outside of her home. She discussed how, for her, the ability to sit and interact within a group in these natural spaces allowed her to begin to address personal difficulties and worries. The group and the outdoor space, the hub site—sat beneath the shade of trees on wooden tree stumps, in a circle surrounded by wood chippings, logs, and greenery—both allowed Holly a sense of safety to tackle what may otherwise be deemed too overwhelming to address when on her own. She is comfortable within this group and tells me she seems to have found her niche in making with wood.

Holly explicitly stated that the difference between being alone and feeling invisible, and being within this group, is in having a sense of clarity and avoiding a mode of escapism where she had difficulty in processing issues. This sense of safety, enabling her to begin to deal with these issues may be due to being in flow,[4] where her skill and the challenge balance in a way as to allow her pleasure. It may also be down to feeling like she *belongs* within this group. For some, the activities allow for a different mode of thought, but for others the simplicity of being within a group is something that they not only desire but allows them to feel more capable of dealing with, and more positive about, their everyday circumstances.

R: Do you think then [...] being within the group makes it better?

G: Oh it does definitely because you can laugh about it and you can always have a moan about it which is just as good as having a laugh [...] that one at Tighnabruaich, I thought that was a bit of a boring job that, but the fact we were all in a group definitely helped that day [...] everybody was in the same field, weren't they? So, you were always mixing with people and interacting with people, that made it OK. Otherwise it would have been quite a boring day that [...] that would have been terrible that... (George)

For Mahar et al.[5] a sense of belonging can be characterised in five elements: (1) Subjectivity, or that it is unique to a person in terms of their feelings of value, fitting in, and being respected. This is not about physically being there or being within a group but about feeling a valuable part of the group[6]; (2) Groundedness: the group is what ultimately provides this subjective feeling,[7] that is the person belongs to something; (3) Reciprocity: this sense of belonging will be felt to be shared by both the person and the group of individuals[8]; (4) Dynamism: the factors that lead to a sense of belonging may be transitory[9]; and finally, (5) Self-determination: a person may feel that they cannot belong to a group regardless of qualifying for group membership, due to physical and environmental factors.[10]

> I was desperate for the group thing. I wanted the camaraderie, I needed [..] I couldn't, I was a member of the gym, I had a cap card,[11] the one-pound card and you'd get up on a Monday morning and go, oh I could cycle to [the] pool and go swimming or I could just stay in bed, and invariably the bed won, do you know what I mean? (Greig)

To suggest that individuals like to belong to a group and that this is their reason for attending these excursions may seem on the surface as a straightforward thing to surmise. However, if we consider the variety of activities, the variety of different individuals, and the multitude of possible motivations for taking part in any activity on a voluntary basis, one would be forgiven for missing this point initially. Given the many variables within these groups, the need to belong becomes a little more complex. It is not about belonging to any group. It is about becoming a part of a group that aligns with one's perceptions of oneself. For my individuals within my case study groups, who all demonstrated a desire to be part of a group, their motivations were often articulated via more practical and less emotional reasoning, with the explicit need for belonging emerging as a by-motivation if you like. If we consider Baumeister and Leary's 'Belongingness Hypothesis,'[12] we are able to consider practical, emotional, and aspirational motivations as one and the same. Regardless of specific motivation, a desire for belonging was at the centre.

For some, the initial joining of the group can have the effect of heightening apprehension regarding how one might be received and perceived. There was a concern that others would not accept them—would they fit in? They worried about who was within the group and about relationships perhaps already formed, as well as about whether others may be better at the activities. Whilst this apprehension sometimes preceded joining the group, for Ross, the interaction was the most positive thing that they get out of joining:

> I think a lot of the things that people are coming in for is that interaction. It's also the thing that scares them the most when they start they're worried […] If it was a female they'd ask how many other females there were because they didn't want to be in a group of lads which is how they thought it might be and they'd ask about, does everybody else in the group know each other?…they wouldn't ask you directly but you'd sense that that's what they're getting at; will everybody else be better at that than me? that's where a lot of them are coming from just that fear of judgement and fear of being rejected and fear of being isolated but yet, that's what they get out of it the most I think. (Ross)

For Holly, although she was aware of feeling isolated as was Greig, she was tentative about joining a group due to her anxieties about what this may entail. Ultimately though it was becoming a part of this group that gave her the incentive to return:

> [..] the whole group thing, for me it's like […] I find it important and it's something that attracted me like as a concept but in reality, it's like oh my god there's loads of people here, like who are these people? do they like me, do I even like them? you know? What is this going to be you know? Am I gonna have to talk about things that I don't want to talk about? Is it going to be small talk is it gonna be like talking about …whatever? all of these things that I think about, that put me off sometimes being in groups and I can be extremely antisocial because sometimes it can really trigger my anxiety and make me kind of have to not be myself […] You know? and it's that whole kind of thing but then you just somehow get through that, or you know I got through that, and then now it's the thing that I really do appreciate the most because the group is the thing that makes me come here […] (Holly)

Holly may never have explicitly stated that she was looking for a group in which to belong; in fact, she may have said quite the opposite, though one might infer due to her exclamation that she appreciated the group, 'the group is the thing that makes [me] come,' that she had discovered something positive about being part of this group. She stated that the group attracted her 'as a concept,' suggesting that the 'group' was an abstract notion in her mind, an idea of something she would like, although did not know if it was something that she would be able to be a part of. Being part of a group may have been outside of Holly's understanding of herself (Fig. 4.5).

These initial insecurities regarding joining a group may come down to a concern over whether one will fit in or self-identify (see similarities in thoughts, morals, beliefs, appearance, and background within the group). In Holly's case, she was not sure if even being one within a group was

Fig. 4.5 Holly weaving a chair

reflective of who she was, and whether she would self-verify (reaffirm one's understanding of themselves) within a group or not. Holly was worried about what may be expected of her, whether she would like the group, whether the group would like her, and how she may be forced to behave. She was concerned about having to perform as someone she was not. She was worried then about fitting in, about behaving appropriately and about whether she would be able to get along in the group. These are signs that Holly was preparing herself for becoming a part of a group of individuals and whether she would self-identify within the group.

Holly lives in cooperative housing and did so for the majority of the time that I knew her, before a fall out with a member of the home community who did not conform with the household ethos. For someone to live within a group, in such a manner, to be unsure of herself within this group struck me as intriguing. This ignited thoughts about the need to belong with like-minded individuals or with individuals who may empathise with one's life choices. She perhaps felt more aligned with the people she chose to live with than with a group who met for specifically mental health reasons, whom she did not know could potentially have similar interests and beliefs as she did. For Holly, then, she was attracted to the activities themselves. She explicitly associates herself with the outdoors, telling me that any time she felt anxious all she needed to do was 'get her bare feet on some grass,' and she would begin to feel calmer. The promise of getting better, along with belonging to a group who were also interested in being outdoors and who were also aiming to get better, provided Holly with a sense of belonging that allowed her to return every week.

For Greig, without having a group of people to meet up with he struggled to get himself out of the house. One could say that his reason for wanting to be part of a group was to ensure that he was, as he states, 'answerable' to a person with whom he had made plans. He felt that he needed a group of others to give him the support to feel motivated. One could also suggest that Greig needed to feel a sense of belonging to feel like he wanted to leave his house. He told me that on some days he barely managed to go to the shops for a pint of milk in his dressing gown and slippers, but when he did, he knew that he felt a little better. The motivation for joining the mental health group then would be to not only feel better and able but to vicariously develop a sense of belonging allowing this change in himself. The group had indeed given him this motivation—a reciprocity. In wanting to be part of a group, he was motivated to join, and in finding what he was looking for within this group, perhaps support, but ultimately a sense that he had a place of significance within this group kept him motivated. Belonging after all is 'characterized by mutual concern, connection, loyalty and trust that personal needs will be met through commitment to the group as a whole.'[13]

> […] I'm better when I'm here […] I do feel better about myself here. If I can make you guys laugh a little bit, I know this is a bit selfish but you know, I feel 'yeaah!' inside you know? but I've always felt like that a little bit, with my colleagues at work and stuff […] I can't speak for anyone else in the room but I've had many a time in my life with complete self-loathing just, 'I'm not the man I want to be,' 'how have I ended up like this, I've made bad choices' and really, self, self-hatred which is quite destructive. I get none of that here. You can lose yourself in a positive environment with nice people who you can relate to a bit. See if you want to have a chat about your mental health situation, most people here will, they've certainly engaged me and there's a lot of lack of having to explain it cos often they've been through it or they've been through something similar and so. My sense of self is quite grounded when I'm here and I feel much happier about me as a man, as a dad, as a friend than I do outside of it […] (Greig)

Perhaps this mutual concern, connection, loyalty, and trust that Chavis[14] characterises as necessities, in the bid for a sense of belonging, can be seen in the above comment. Greig felt his place within the group. He spoke of making the group laugh, a reference he made often. I got the impression that Greig perceived his role as in the light-hearted cheering of others. He had been the class clown, or 'office clown' as he put it, for the majority of his life and in this role, he felt himself. Greig seemed able to converse with anyone. He always had time to ask how people were and to provide an ear for listening. The group joke was that he always broke the silence, and he sometimes worried about whether this was appropriate. Before becoming a part of this group, Greig felt self-loathing; he explicitly stated that he had often thought that he was 'not the man [he] want[ed] to be.' Now, within this group, in this 'positive environment,' apparently in stark contrast to his previous working life, he felt not only as though he was relating with people who could empathise with his situation but that he was 'grounded.' This allowed him to feel happier about himself as a man, a dad, and a friend. A sense of belonging is 'highly shaped by specific contexts [and] generally involves a sense of connectedness, positive interaction with social others, and a highly complex performance of identity.'[15] Here Greig was able to perform more closely in line with his ideal sense of self.

Jess found her belief, prior to joining the group, that people may empathise with her mental illness situation was a precursor to being able to identify with and be herself within the group. Her guard was perhaps not as ready as it may have been had she not pre-empted this empathic scenario:

> I think it helped me initially that it was a project for mental health because I kind of thought, because I was very worried how like, if I was behaving strangely and stuff, if I was having a bad day I thought right well it's a probably for mental health, people know that we've all got some condition or other or whatever so [...] (Jess)

Jess was concerned about 'behaving strangely,' something that she was often concerned about in other public and social situations. She felt, here however, without the need to specifically state that she suffered with mental health issues and that this may affect her behaviour if she was 'having a bad day,' that people would accept her regardless. Knowing that others within the group may have similar issues in advance allowed Jess to be more open to the group, laying the ground for an almost automatic sense that she belonged here. This was an identification with the group and with her own sense of self. Within a mental health setting one wonders whether a knowledge that others may empathise creates an automatic in-group mentality, where collective 'problems' and empathy merge in 'a continuum rather than another plane.'[16] I would suggest that this allowed for a more quickly formed sense of cohesion amongst this group of individuals:

> There's a lot of lack of having to explain it cos often they've been through it or they've been through something similar [...] (Greig)

If we consider Forsyth[17] on social facilitation, perhaps the set-up of this group—facilitated by the organisations staff, via public health bodies, with casual minimally structured group meetings—allowed for 'dominant responses' to both the activity and behaviours within the group. By dominant responses, Forsyth suggests 'easy responses.'[18] Social facilitation—perhaps the creation of this safe space—made for an easier interaction amongst the group. This is in contrast to what Forsyth considers to be 'social interference,'—a scenario requiring 'untried behaviours' that would potentially inhibit performance. 'Social interference' is associated with tasks and behaviours that may then be better performed in isolation[19]—a common prevention of some of the group engaging in other social situations however not found here. Forsyth was talking of shared tasks. This concept could be expanded to encompass, in this case, the simple, yet challenging for some, act of being within a group. The dominant responses then become the natural and easy way to behave in comfort, as oneself as opposed to attempting untried behaviours in order to fit into the group.

The way that the organisation pitches itself to participants offers a potential to be at ease within a group scenario that may otherwise not present itself to these individuals. Perhaps then individuals are more comfortable with their own vulnerability? Within this group, actual diagnoses were rarely spoken about. Jess stated herself that they didn't speak about specific issues but that she would have been more than happy to if people had wanted to. This is a sentiment that was echoed by both Greig and Holly. There was an unwritten understanding, a social contract of sorts, that you just don't pry. The knowledge that people are likely going through something similar to others within this group allowed for a predetermined 'fellow-feeling'[20]—an imaginative empathic feeling amongst the group. These people had already made the decision to take part in something that offered to help in feeling better. They had already begun a personal journey and had met others along the way. Rather than perhaps being like-minded, the participants had found themselves in similar circumstances to others within the group. This was similar to the young adults who may not have had much in common with others within their group besides a need for support in external life factors. This distinguished these two case studies from The Loose Community and The

Community Living Initiative in that these people, rather than being brought together due to similar interests or morals were brought together by a desire to get better or to improve oneself.

Baumeister and Leary[21] specify that changes in one's 'belongingness status' will result in an emotional response. In order to feel this sense of belonging, an individual needs to 'believe that the other cares about his or her welfare and likes (or loves) him or her'[22] and the reason behind the need to belong is twofold: (1) 'People require frequent interactions with the same person,' and (2) 'people want a stable, enduring context of concern and caring.'[23] Within Maslow's hierarchy of needs, 'love and belongingness' are also deemed important. They do not, however, place in this hierarchy before the basic needs are met: 'Love and belonging does however come before esteem and self-actualisation.'[24] Self-actualisation may come later when the self develops more closely in line with their ideal or ought self.

Below Ross compares this desire to belong to more complex understandings of self-identification and being attracted to the idea of a group and belonging to a group:

> [...] in a way, because were all massively tribal and we're all kind of instinctively trying to fit into this tribe or that tribe and to belong, that's basically where like football, like I don't find football very interesting but that's where that whole football thing comes from, you know being so proud of a team. I don't really know why but to just say you know, I am this. If I say I'm this you're by association saying I'm not that, that's what, that's why wars start but it's also where our sense of belonging comes from so I think you need to, something needs to draw you in, I don't think you could do it in a vacuum. (Ross)

For Ross, the need to belong is accompanied with a need to identify with others, a need to find similar interests and to be able to associate oneself with an external factor—to self-verify or satisfy an extended sense of self.[25] Ross described the need to belong as tribal. It is to do with wanting to be part of a distinct group that we feel represents us in some way. He compares joining a group to supporting a football team. For many, being able to say that they are a specific team supporter says something about them as a person. The team that they support is a part of who they are and distinguishes them from who they are not. For Ross, this is the same sentiment as joining any group. We seek to say something about who we are and to distinguish ourselves from other groups of individuals. This in-group mentality is quickly formed when individuals are able to make this distinction. We are drawn to groups that offer the opportunity to belong but also to offer us a means by which we may define ourselves. In this, Ross suggests that those who join both The Mental Health Initiative and The Woodland Weekend Group see something within the groups that they would like to associate with themselves and in doing so find a sense of belonging within the group.

Baumeister and Leary's 'belongingness hypothesis'[26] argues that it is a basic human instinct to desire to belong to a group. They believe that all actions within social life are motivated by this ultimate desire to belong. The 'belongingness hypothesis' would suggest that the natural landscapes attended in these excursions, though significant, play only a part in both motivational calling and the situations themselves as personally and socially transformational. One may seek to belong to a certain kind of people, to identify oneself with certain kinds of activities or modes of thought or indeed, to identify with certain types of landscape.

Belk's concept of the 'extended self' intimates that activities, particular places, experiences, belief systems, behaviours, types of people, and so on might be considered to represent who one is or who one would like to be. This could be in the arts, sportsmanship, volunteering or altruism, life-long education, or spirituality, and so on, and there have often been associations made with the types of activity carried out to one's personality and motivation. Perhaps it is not surprising that people wish to partake in activities that interest them with people who interest them.

Ross believes that an important aspect of this for many of the people who attend the woodland weekends, and for those who become *regulars*, as is often the case, is the ability to revisit places:

> [This] feeds into, kind of, it's not primal it's not as ominous as that but this, this desire that people, that a lot of people, I certainly have it, to have not ownership but stewardship over a piece of land.

It becomes then, not only about leaving the urban behind but in a sense staking some claim or placing some aspect of your identity in the natural landscapes in which we encounter. For Ross, this became more obvious when the group were able to see the development of the land whilst working within it and to associate themselves with that physical transformation that was thought of as positive.

> I've probably spent on and off maybe two years just sort of living rough and travelling through, it was not always a natural environment but mostly in natural environments and yeah, I intentionally, though slightly unconsciously at the time though intentionally did that in order to process something, to understand something and to go through something and yeah, being away from an urban environment was pretty key to that. (Ryan)

Ryan, a member of The Loose Community, has devoted two years of his life and career to 'processing something' and to 'understanding something' and 'experiencing something' within nature. Again, we encounter that ineffable *something*. In Ryan's context, this refers to an ability to communicate with and to understand the natural world. Whilst this seems in contrast to George's other-centred reasoning for taking part in excursions, I don't believe that these motivations are so different. Ryan runs a community project and intends to utilise the knowledge and understanding that he gains within the underprivileged groups in which he works to process, experience, and understand this *something* within the natural landscape. Ryan insists that being away from an urban environment may be key. He suggests that at the beginning of this process he may not have in fact been doing anything consciously; however, he did approach excursions intentionally. By 'intentionally,' he means, with reference to mindfulness, that although he may not have been fully aware of his specific aims or indeed specifically what the *something* that motivated him was. He was aware that in order to discover it, he needed to be within the natural landscape with intention and openness. His motivation then was a metaphysical one—to feel a sense of place within nature.

> I pinpointed that creativity is really important but not just general creativity, not just anything but what was missing was the doing something with

> my hands […] creating something that didn't exist before and improving something and when it's finished you get that sense of oh I […] I knew that I needed to do that. […] in that like really dark period I was going through […] I just ran with it and I just sort of held on to it and said I'm doing that […] (Holly)

Holly associates herself with creativity. In order to make herself better, she sought activities that she felt defined her as a person. She found activities that would help her to understand herself. Holly believed that this was to do with using her hands. She needed to do something tangible to feel that she was making progress in herself. This is an example of how the object that she made might be considered as an extension of herself. The tangible object, that was not perishable, not edible, not temporal, but solid, would become the extension of who she was. She wished to create 'something that didn't exist before.' Holly needed to be able to physically appreciate what she had achieved. In a particularly 'dark' period of her life, she found solace in finding something creative to do. This would allow an external object to transpire upon which she may be able to reflect and project her sense of self. Holly's motivation then, was not simply to be part of a group as mentioned before but to physically be able to demonstrate a positively transformed sense of self through her ability to make objects.

> 95% of the places you go are community woodlands so we, you're not just getting stuff for yourself out of it, you're getting, you're helping the local community which is always a good feeling […] I know the ones [local community members] we do speak to are very grateful for what we do because even if they're with the best of intentions, these people don't have the man power or the time to do big jobs. […] We can go in for a weekend and do what would take them two or three months […] (George)

For George, an altruistic person, his motivations may appear to lie partly in doing for others. He enjoys helping the local community, regardless of the fact that we very seldom met these people. This led to a 'good feeling.' He spoke of the local people being grateful because they wouldn't have been able to 'get the work done' otherwise. They were grateful for the work that George and The Woodland Weekend Group did, and for George this was positive.

Eisenberg et al.[27] define pro-social behaviour as voluntary behaviour for the benefit of others. We might say then that George is a pro-social person, and this is what drives him. It may also be why he gains satisfaction from such excursions. The desire to behave in a pro-social manner may in fact be motivated by many things—including the desire for reciprocity and social approval.[28] This may tie closely with a desire to perform one's identity as one aligned with altruism. These kinds of behaviours also reflect moral values for example relating the equality of people or a care for others.[29] Although George gains by way of positive social outcomes with regard to these weekends, ultimately, he morally aligns his sense of self with the values of social responsibility for others.

The Ideal Sense of Self

George and Holly take part in these excursions to satisfy an element of their own perceived ideal sense of self. Their ideal sense of self is a projection further than their actual self. By taking part in these journeys within the groups, they found a sense of positive wellbeing due to fulfilling an aspect of their identity that they are not able to on a day-to-day basis:

> I think we're all a little bit insecure about ourselves and we're all actors and actresses and we all wanna, life would be great if we were on a stage all the time being loved, wouldn't we, we wanna be loved! We're self-interested animals, aren't we? We want to be loved [...] if I was in a secluded part of the highlands doing this work and there was no phone reception and I was by myself and I was doing a lot of work you know, I'd be like, the logical reaction to that would be like, shit you've done a good job, you've done a lot of work but for me it would be like 'yeah I've done a lot of work but nobody saw it!' [Laughter] (Navid)

According to Higgins,[30] there are three domains of the self: the actual, the ideal, and the ought self. In other words, he refers to (a) the way we are, or the way that we or others represent our own attributes (actual self), (b) the way we would like to be, or the way others would like us to be (ideal self), and (c) the way we think we should be, or the way oneself or others believe us to have the duty to be (ought self).[31] Additionally, Higgins refers to Freud's 'superego' and 'ego ideal.' These can also be referred to as 'personally relevant self-guides.'[32]

> Possible selves are *the ideal selves* that we would very much like to become. They are also the selves we could become, and the selves we are afraid of becoming.[33]

I consider here also the 'possible self.'[34] This is a further concept of self-knowledge with regard to personal potential and links cognition and motivation.[35] The possible self is future orientated. Though nobody explicitly spoke of their 'future selves,' they did speak of their hopes, fears, goals, and aspirations. For Markus and Nurius,[36] it is this dynamics of the self-concept that leads to motivation (as well as distortions of the self). This could be said to be implicit in individuals' desires to take part in these activities. The 'possible self' is the 'self to be approached or avoided.'[37] Members of the groups, by taking part, implicitly approached a sense of self that they aimed to be. This sense of self could be 'the creative self,' 'the successful self,' 'the admired self,' or 'the loved self.'[38] If we consider push and pull factors and individual motivations to take part, they may be to do with both positive ideas of the future self and negative ideas of what one does not want to become.

Though a fear of whom one might become was never explicitly articulated, it may well be appropriate to presume that if one was taking part with the objective of positive goals one would be taking part to prevent negative perceptions of the self. If, as Markus and Nurius state, the concept of one's 'possible self' often 'give[s] meaning to current behaviour,'[39] then joining these groups relates specifically to a context of one's own self-knowledge, and this is often based on social comparison.[40]

For Kyle et al.[41] identity is the primary motivator of an individual's behaviours. James[42] further identifies the 'spiritual self' incorporating one's morals and conscience. He also identifies the 'social self' incorporating the self that may or may not be 'approved' by 'the highest social judge.'[43] One might gather this as the hypothetical or metaphorical imagined adjudicator of one's life choices and behaviours, for some, religious, for others not so. The source of the beliefs regarding the ought self and the ideal self or how one should or would like to be comes often from what Turner[44] refers to as the 'normative reference group.' This could be described as one's peers. There is additionally the 'extended self' as mentioned previously: 'The body, internal processes, ideas, and experiences, and those persons, places, and things to which one feels attached.'[45] This 'extended self' is an association of external factors as attainable within ones understanding of the self:

> A man's Self is the sum total of all that he CAN call his, not only his body and his psychic powers, but his clothes and his house, his wife and children, his ancestors and friends, his reputation and works, his lands, and yacht and bank-account. All these things give him the same emotions. If they wax and prosper, he feels triumphant; if they dwindle and die away, he feels cast down, -not necessarily in the same degree for each thing, but in much the same way.[46]

> People these days are fond of pointing out that you are what you eat. That proposition is true enough, but there is another which I think is a good deal more profound, namely, that you are the company you keep. Your identity, your self, depends upon the people and things that compose your associations. And perhaps even more important, your knowledge of yourself and your development as a person are both predicated on those same associations.[47]

These are two interesting ideas—the self is everything that one might call one's own *and* the self and is dependent on the company one keeps and one's associations. If so, gaining self-verification and closing the gap between the actual and ideal self may be understood to be down to *what* you have (perhaps sometimes experiences and sometimes places) and *who* you have (those who you feel may be inspirational perhaps, or whom you might look up to or deem good). I believe this is a key thing to comprehend when thinking about why someone might choose to become a part of a specific group in order to make these excursions.

By introducing the concept of the 'extended self,' I further develop ideas that individuals within groups may hope to not only find a sense of belonging but to self-verify by becoming part of a group. By considering their ought or ideal self, they may perceive certain objects, places, and activities, ways of dressing, or perceived kinds of people as being the kind of people or activities that they should spend time with or doing. This is found in Belk's notion of the 'extended self.' In this, he includes external objects, people, and places. When people claim objects as 'mine,' they simultaneously come to believe that the object is 'me.'[48] This may also refer to one's experiences: the experience was mine, the experience was me. This thought process directly links experience to one's personal narrative. We might see this in relation to kit, tools, and associated objects to being outdoors: this tent is mine, camping is me, for example. I believe more importantly we see this in people's relationships to the group and to the outdoor environments: these people are mine, this landscape is mine—*they are me*. The groups of people with which one associates and the natural landscapes in which one engages becomes closely linked to not only one's personal narrative but their sense of who they are as a person. By associating with, deemed like-minded people in self-congruent places one's ought or ideal self may be believed to be coming closer to the actual self. This is similar to the '*ecological self*,' the self that incorporates all that we identify with.[49]

Self-Identification and Becoming a Part of the Group in Landscape

I now explicitly link the concept of self-verification and the ideal sense of self with the desire to belong. Ultimately, I link this to reasoning as to why shared experiences in natural landscapes may be deemed positively transformative. I will fully expose the importance of these concepts within the findings of my fieldwork. I will do this by analysing one informant and the comments he made to me whilst walking and in post-excursion conversation and relating these to similar themes and comments made by others.

Navid considers himself as an 'outdoorsy, Bear Grylls kinda guy.' This is apparently at odds with his other 'persona' as an actor and appreciator of fine art and 'European avant-garde cinema.' He attributes this to an 'insecure and masculine thing,' and he believes that there are a lot of men in Britain, but in Scotland specifically, who choose to flaunt their masculinity. For Navid, these men don't necessarily want to nor do they need to. It is simply a cultural gender stereotype that encourages this behaviour. Amongst this kind of behaviour, Navid says, are activities such as being outdoors, setting up camp, sleeping in the 'wild,' cutting down plants, and climbing trees. It is this association with masculinity that Navid feels entices men into doing activities such as the ones within the woodland weekends:

> [...] a lot of guys would be attracted to this, not necessarily cos they agree with the work but more because of the nature of the work. It fits in with their masculinity.

For Navid, this is one of his reasons for attending too:

> I'm probably interested in art more than I'm interested in outdoors work but the reason I do more outdoors work is because of the image I want to create about myself, maybe I'm quite insecure about myself? so I wanna go back and tell all my mates like 'aw you know what, look where I was, look where I was sleeping, sleeping in a [poly] tunnel' you know, even though I didn't enjoy sleeping in the tunnel, my mates are like 'aww you're the man for sleeping out there' [laughter] and I'd be like 'yeah cheers mate' you know?

Navid attended an all-boys school and at around the age of 16 suffered with anorexia. He told me that he has always had body-related issues and due to this had 'issues about [him]self [and] issues about [his] identity.' As a British Indian, Navid always felt as though growing up in this male-dominated environment 'wasn't good for him psychologically.' He believes this was because he grew up adhering to two conflicting cultures: Indian and British. This has left him feeling like he does not 'know his place in society,' leaving him self-conscious about his own masculinity:

> If I'm working and stuff, it may sound weird, but I would work harder if I had an audience watching me.

Navid explicitly admits to performing as an aspect of his identity. He does this in order to reflect an aspect of himself that he is insecure about to the people that he is around. He takes part in these activities to self-verify the aspect of himself that he believes should behave in a certain way and have certain attributes. In conversation with Craig, Navid came to the conclusion that he too attended the woodland weekends due to being bullied at school. He felt that several people within the group at some point in their lives lost confidence and no longer felt able to do 'regular' things. Perhaps in a similar way to the individuals within The Mental Health Initiative, public and social situations have become a little harder for some within The Woodland Weekend Group. Attending these trips became about proving something to 'the a**holes that treated them like s**t, [showing them] that they're doing all this amazing stuff.'

On every trip, Navid took and posted a picture of himself on Facebook and Twitter. This was likely to be either a photograph of himself chopping wood, standing by his tent, or looking over a mountain. He admitted that this was unnecessary—after all, he could have simply called his friends and family and told them about the trip. He told me that he had photographed himself outside of his tent and felt good when he received positive feedback on these social media sites. He 'felt good about [him] self'—he self-verified an aspect of his perceived identity. He admitted that it may have made more sense to photograph the whole group and that this would have provided a more accurate representation of the weekends. It would not however present the aspect of his identity that he wanted to. Navid is not alone in this behaviour as I witnessed it often. On a wood fuel collection day, Victoria, a bar tender and clinical support worker, asked me several times to take her photograph. She exclaimed: 'I do outdoorsy stuff too!' She wanted to send photographs to her partner, and when she did, she followed it with the text: 'Get out your bed, look what I've been doing since 9 am!' She was then frustrated when he failed to provide the response that she wanted from him. When we were finished for the day, she showed me all the photographs that she had taken, asking for reassurance that she 'looked the part.'

For Beech, 'the social identity consists of projections of others towards the self, projections of the self towards others and reactions to received projections.'[50] For Twigger-Ross and Uzzell,[51] 'there are four principles of identity which guide action' and these are 'continuity, self-esteem, self-efficacy, and distinctiveness.'[52] Identity work literature focuses on identity as being dynamically constructed and reconstructed.[53] Beech discusses the ideas of how the identity is so: we are cast by others, we project an identity, and we take on behaviours and stories. He reaffirms this when speaking of the social identity's influence on the self-identity.[54] Ironically,

> we behave towards the other in terms of our understanding of what we guess are their beliefs and desires rather than directly in terms of what they appear to be like externally. The complexity of human social life is thus built on this continual imagination of the minds of others. This is a process of ever increasing complexity since we act towards others not only in terms of what we believe others believe and desire but inevitably also in terms of what we believe others believe are our beliefs and desires, a process which can go on and on [...][55]

Navid comes from a military family, his grandfather having fought for the Indian air force. He becomes frustrated when members of The Woodland Weekend Group suggest that those that have fought in wars are bad people. He is defensive about this because his grandfather is his 'hero.' This is a typical response from a group of people who champion pacifism and a group who openly discuss left-wing politics. Navid sees himself as a little more right wing than the 'liberal kind of agenda which is in [the group].' Navid second guessed the beliefs of the group. For Navid then, as he did not necessarily identify with all of the group's opinions or politics found sense of his self-verification in the activities and the locations. For others, who do align themselves with the group's ideals, they will self-verify with the group also:

> I think the key thing that you've said about like, what does it make, how do we feel about our sense of self it, there's pride, in the positive sense and I think everyone's felt proud of something that they've done and that grounded-ness as well, I think someone else said that they feel grounded, I think you said that [to another participant] when they're here and I feel like that as well and I feel also not just grounded and not just proud let's say but comfortable and relaxed with feeling those things. (Holly)

The desire to self-verify the ideal or ought self was evident in participation. For Holly, from The Mental Health Initiative, this manifested in feeling a sense of pride at having worked on something that she aspired to do. She had pride in having had an experience that she could now associate with her own identity. This could be in making something, or visiting somewhere, or taking part in a group activity. This then became something that she was now able to do and thus was a part of her identity. For Holly, this came with a sense of being comfortable in feeling emotions that she was not familiar with. Previous to these experiences she had not associated pride nor groundedness with who she was.

For Patrick, he felt his whole self when he was outdoors:

> I do feel myself, to the greatest extent when I'm out in that environment really so I feel happier there than anywhere else and it's doubly good if you're with other people who enjoy being out in the natural world and if I can be working with them and showing them how to do things you know that's fantastic, I get a lot of job satisfaction out of that. (Patrick)

Patrick enjoyed being outdoors due to feeling himself 'to the greatest extent.' It was not simply the outdoors that led him to feel more himself—it may also be to do with sharing this space with like-minded people who verify his way of life and thinking. Additionally, as he taught the groups, he felt a sense of purpose within this situation. Burke and Stets[56] believe that a person's identity will be verified through interactions, shared experiences, and locations, and it is because of this that, through seeing them as dependable, an individual will begin to feel 'confident and secure in the likelihood of self-verification of their identity.'[57]

The figure labelled 'Relationships of Interest' (Fig. 2.2, in Chap. 2) shows how I believe the three subjects of self, group others, and place may interact. We will return to this. Now that I have added this further dimension, it may also be imagined a little more like the larger figure around it. Whilst we have the self, group, and landscape all relational to one another in terms of how they aid in transformation, below we also see how the ought self or the ideal self influences the decision to become part of a group. It then shows what effect joining these groups has on informants (Fig. 4.6).

I propose the above figure as a means by which to understand the relationship between the motivation towards embodying the ought or ideal self, self-verification and a sense of belonging, and a sense of wellbeing.

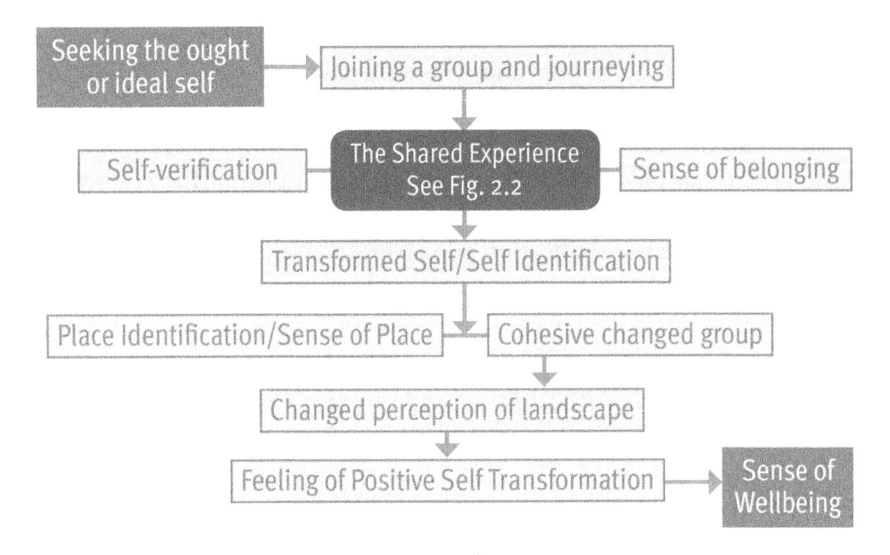

Fig. 4.6 Motivation to sense of wellbeing via shared experience of natural space

One has a belief in what kind of person one would like to be—this can be described as either the ought or the ideal sense of self. The ought self is that which motivates and guides us or that which we believe others would like us to be. The ideal self is that which we would like to be.[58] These are both separate from the actual self, though attainable through action, thought, and behaviour. In seeing affinities with a group or in seeing something that they identify with or believe they will get from being part of a group (friendship, inspiration, new skills, improved mental health, or whatever that may be), the individuals with whom I researched sought to become a part of their respective groups.

The Young Adults sought to change something within themselves and identified the personal development programme as somewhere where they might achieve this. In the case of The Mental Health Initiative, they may also see the group as something that may help them. They may join this group due to the understanding that there will be other people within the group who also suffer from mental health issues. The Woodland Weekenders, The Community Living Initiative, and The Loose Community may see elements of the person they would like to be within the morals of the group, the beliefs of the group, or similar interests within the group. All of them join a group and through being within this group, a sense of belonging and self-verification manifests within the time spent with these people in these landscapes. This often led to a witnessed change in the person. What followed was a cohesive and bonded group, a sense of place and thus a changed perception of the landscape, and how the landscape offers opportunity. Through achieving a sense of belonging and self-identification alongside a sense of place and this bonded feeling with the group, a person will articulate feelings of positive self-transformation. They articulate a sense of feeling better. This varies from person to person in terms of what feeling better means to them. We might say, though, that they find a sense of wellbeing.

Notes

1. Baumeister and Leary (1995: 497).
2. Turner (1969, 1970, 1974).
3. Jones (2011: 160).
4. Csikszentmihalyi (2002).
5. Mahar et al. (2013: 1030).
6. Mahar et al. (2013).
7. Mahar et al. (2013).
8. Mahar et al. (2013).
9. Mahar et al. (2013).
10. Mahar et al. (2013).
11. A 'cap card' is an Edinburgh City Council leisure card, allowing access to gyms, pools, and other leisure activities for only £1. These cards are given to those with low income, or through GP prescription.
12. Baumeister and Leary (1995).
13. in Mahar et al. (2013: 1028).
14. in Mahar et al. (2013).
15. Caxaj in Mahar et al. (2013: 1028).
16. Malekoff (2010).
17. Forsyth (2013).
18. Forsyth (2013).
19. Forsyth (2013).
20. Adam Smith (1759: 2010).

21. Baumeister and Leary (1995).
22. Baumeister and Leary (1995: 500).
23. Baumeister and Leary (1995: 511).
24. Maslow, in Baumeister and Leary (1995: 497).
25. Belk (1988).
26. Baumeister and Leary (1995).
27. Eisenberg et al. (2010).
28. Eisenberg et al. (2010).
29. Eisenberg et al. (2010).
30. Higgins (1987).
31. Higgins (1987: 320–321).
32. Higgins (1987).
33. Markus and Nurius (1986: 854, emphasis my own).
34. Markus and Nurius (1986).
35. Markus and Nurius (1986).
36. Markus Nurius (1986: 854).
37. Markus and Nurius (1986: 945).
38. Markus and Nurius (1986: 945).
39. Markus and Nurius (1986: 955).
40. Markus and Nurius (1986).
41. Kyle et al. (2014).
42. James in Belk (1988).
43. In Higgins (1987).
44. Turner (1969).
45. Belk (1988: 141).
46. William James in Belk (1988: 139).
47. Gregory Bateson in Belk (1988: 156).
48. Belk (1988).
49. Naess (1987).
50. In Beech (2011: 286).
51. Twigger-Ross and Uzzell (1996).
52. Twigger-Ross Uzzell (1996: 205).
53. Twigger-Ross and Uzzell (1996).
54. Twigger-Ross and Uzzell (1996).
55. Bloch (2012: 63).
56. Burke and Stets (1999).
57. In Kyle et al. (2014: 1025).
58. Higgins (1987).

References

Baumeister, R. F., & Leary, M. R. (1995). The Need to Belong: Desire for Interpersonal Attachments as a Fundamental Human Motivation. *Psychological Bulletin, 117*(3), 497–529.

Beech, N. (2011). Liminality and the Practices of Identity Reconstruction. *Human Relations, 64*(2), 285–302. https://doi.org/10.1177/0018726710371235.

Belk, R. W. (1988). Possessions and the Extended Self. *Journal of Consumer Research, 15*(2), 139–168. Oxford University Press. Stable URL: http://www.jstor.org/stable/2489522

Bloch, M. (2012). *Anthropology and the Cognitive Challenge.* Cambridge: Cambridge University Press.

Burke, P. J., & Stets, J. E. (1999). Trust and Commitment in an Identity Verification Context. *Social Psychology Quarterly, 62*, 347–366.

Csikszentmihalyi, M. (2002). *Flow: The Psychology of Optimal Experience.* London: Rider/Random House.

Eisenberg, N., Eggum, N. D., & Di Giunta, L. (2010). Empathy-Related Responding: Associations with Prosocial Behaviour, Aggression, and Intergroup Relations. *Social Issues and Policy Review, 4*(1), 143–180.

Forsyth, D. R. (2013). *Group Dynamics* (6th ed.). Pacific Grove: Brooks/Cole.

Higgins, E. T. (1987). Self-Discrepancy: A Theory Relating Self and Affect. *Psychological Review, 94*(3), 319–340. Available at persweb.wabash.edu/facstaff/hortonr/articles%20for%20class/Higgins.pdf. Accessed 14 June 2018.

Jones, O. (2011). Chapter 11: Materiality and Identity – Forests, Trees and Senses of Belonging. (Editors Proof). In E. Ritter & D. Dauksta (Eds.), *New Perspectives on People and Forests, World Forests* (Vol. 9). https://doi.org/10.1007/978-94-007-1150-1_11.

Kyle, G. T., Jun, J., & Absher, J. D. (2014). Repositioning Identity in Conceptualisations of Human-Place Bonding. *Environment and Behaviour, 46*(8), 1018–1043. https://doi.org/10.1177/0013916513488783.

Mahar, A., Cobigo, V., & Stuart, H. (2013). Perspectives in Rehabilitation: Conceptualising Belonging. *Disability and Rehabilitation, 35*(11–13), 1026–1032. https://doi.org/10.3109/09638288.2012.717584.

Malekoff, A. (2010). Difference, Acceptance and Belonging. *Social Work with Groups, 14*(1), 105–112. https://doi.org/10.1300/J009v14n0109.

Markus, H., & Nurius, P. (1986). Possible Selves. *American Psychologist, 41*(9), 954–969.

Naess, A. (1987, August). Self-Realization: An Ecological Approach to Being in the World. *The Trumpeter: Voices from the Canadian Ecopsychology Network, 4*(3), Victoria. Available at http://trumpeter.athabascau.ca/index.php/trumpet/article/viewFile/623/992. Viewed 27 July 2017.

Smith, A. (1759/2007). *The Theory of Modern Sentiments*. New York: Cosimo, Inc.

Turner, V. (1969). *The Ritual Process – Structure and Anti-structure*. Suffolk: Richard Clay (The Chaucer Press Ltd).

Turner, V. (1970). *The Forest of Symbols: Aspects of Ndembu Ritual*. Ithaca: Cornell University Press.

Turner, V. (1974). *Dramas, Fields, and Metaphors*. Ithaca: Cornell University Press.

Twigger-Ross, C. L., & Uzzell, D. L. (1996). Place and Identity Processes. *Journal of Environmental Psychology, 16*, 205–220.

5

The Liminal Loop

Liminality sits at the edges and intersections of these relationships. This is *the liminal self* in encounter. I am referring to the individual, intrapersonal changes, transformations, and inner emotional movements that occur throughout encounters towards and within the group, towards and within the landscape, and in thoughts about the self in context. These are transformed ways of thinking that may sustain in daily life post-experience. When I refer to the liminal group, I refer to the changes within the group that may or may not lead to group contention and the changes in collective perception of the landscape and activity within and the emergence of 'communitas.' The *liminal landscape* refers to my concept that for the groups and individuals observed, the landscape changes in terms of meanings attached, value attributed, and so on; the perception of the landscape is liminal. This may be found in the development of personal contexts. The word *transformation* at the core of the diagram below, and at the core of this book, refers directly to my finding that due to these encounters and the liminal boundaries crossed (of being 'betwixt and between'[1] these metaphorical and literal spaces), there are articulated feelings of transformation.

Each space marked 'affect' is a potential liminal boundary that metaphorically sits between the self and the group, the group and the landscape,

R. Crowther, *Wellbeing and Self-Transformation in Natural Landscapes*,
https://doi.org/10.1007/978-3-319-97673-0_5

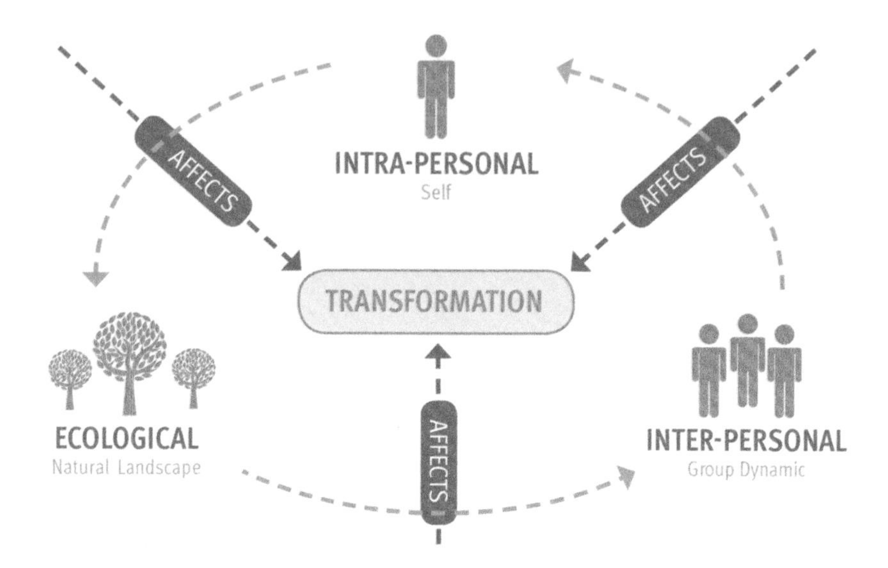

Fig. 5.1 Relationships of interest—nature: culture

and the landscape and the self. These are the metaphorical thresholds sitting in the mind and body, the group dynamics, and the perception of the landscape. This is the focus of this chapter (Fig. 5.1).

When I say liminal, I refer in part to van Gennep's[2] work on liminal rites of passage; however, more so, I am borrowing from Turner's[3] furthering of Gennep's theories. For Gennep, the liminal spoke of demarcated sacred space and a quality of time, separate from the profane and the secular,[4] or a 'cultural realm "out of time".'[5] These are the transitional spaces found within the 'margins,' the 'limens'—in other words, the thresholds. These thresholds for van Gennep were liminal, with the potential for social and personal transformations.

From urban to natural, from motorway to dirt track, from land to water, and from pathway to forest—when people crossed physical boundaries, these boundaries could be said to have been liminal thresholds. They were liminal in the sense that they marked between familiar and unfamiliar landscapes. In this, there was a sense of ambiguity of crossing over with a moment of being between lands. Thresholds initially presented themselves in the field as literal. It soon became evident, though, that liminality could be considered as metaphorical and related to the self.

There was clear movement through thresholds. These groups made a conscious effort to leave urban environments in pursuit of natural ones. They were in pursuit of *different from the everyday* experiences; they chose to separate themselves from their urbanity. Groups did this to experience *something* that they believed was synonymous with this kind of activity, and they returned to their everyday post-excursion. The excursions themselves, in the way that they are framed by facilitators, participants, and practitioners, are set up as separate, or to use van Gennep's term, 'demarcated' from the everyday activity of the groups. These activities were new, ambiguous, temporal, and different in structure to the groups' earlier day-to-day activities.

Turner[6] furthered van Gennep's ideas by stating that perhaps modern, post-industrial society was instead 'liminoid,'[7] where liminal experiences could be found in not only ritual ceremonies but in profane leisure activities and in play.[8]

> Liminoid is felt to be freer than the liminal, a matter of choice rather than obligation.[9]

The transition stage for Turner[10] was a kind of 'social limbo' prior to returning to a new position in society. However, he suggests that these 'may also include subversive and ludic [of the nature of play] events,'[11] where there could be a 'dissolution of normative structure,'[12] an anti-structure or subversion of norms. This anti-structure, for Turner, relates to the delineation between work and play or leisure, 'which includes but exceeds play.'[13]

These moments and activities that Turner refers to as 'liminoid' 'resemb[led], without being identical with, liminal.'[14] This is a temporal space of which he refers to simply as 'free -time,' or 'neutral space' and 'an independent domain of creative activity.'[15] This time, unlike that of van Gennep's ritual encounters, was down to individual choice to take part. It was no longer only thought of as ritual obligation.[16] Instead, Turner considered the liminal as a 'betwixt and between' the norms of work and civic or familial activity.[17] It accompanied a 'freedom from' and 'freedom to' do as one liked. It encompassed a freedom 'to play,' 'to transcend social structural limitations,' and perhaps more fittingly to this research, 'a chance to recuperate and enjoy natural biological rhythms again.'[18]

The liminoid became clear in the activities and journeys of my case study groups. They were not necessarily sacred but instead encompassed time taken away from the everyday. These excursions were demarcated by the leaving of the urban. People were separated from their normal day-to-day commitments and routines. Within these moments, it became clear that *something* intangible did in fact happen. Whether it was physical or metaphorical liminal thresholds that were observed, neither were necessarily dogmatically sacred nor profane. These thresholds came with personal transformations and new ways of thinking and speaking about experiences with the groups and landscapes.

I borrow Turner's notion of the liminoid[19] with regard to where 'the liminal' may be found in social, 'ludic' space, 'free-time,' and leisure activity within natural environments. Turner allows me to explore the notion of liminality within these kinds of excursions. The liminal was found in this mundane social interaction. I adopt the term 'liminal' from van Gennep, paying heed to the specifically ambiguous, transformative nature of these excursions.

Making *Special*: The Framing of Activity

Here I return to the notion of performance frames as discussed in Chap. 2. Key within all case study groups is that each activity, whether mundane, pragmatic, therapeutic, or spiritual, was made *special* by the facilitators and by the set-up of the excursions. Often excursions were set up to inform a certain kind of experience. For example, The Young Adults' excursions were set up as adventures or for personal development. Activity was often painted as something 'normal' one might do if one was functioning 'well' or to the best of one's abilities. (It is suggested that this may be in the same way as those who do not have difficulties might do.) In this instance, a lot of talk is about contribution, finding work, and 'living well.'

For The Loose Community, the artistic and shamanic activity was geared towards focusing attention or enchanting the landscape and slowing down. The individuals enchanted the mundane. They suggested something deeper and indicative of bettering one's 'spiritual' self.

Some participants visited these locations in order to provide a kind of sanctity or respite from difficult contexts. This was intended to allow physical and metaphorical space to challenge oneself. Some were voluntarily attending such activities to do some hard work and to socialise. The perceptible benefits to do with wellbeing or transformation were considered as by-products to this. There were also those who sought this 'framed' space to connect with themselves and with the natural world.

The journeys as well as the activities and situations that took place once we reached our destination were liminal in the sense that these experiences involved a before, during, and after. Much like the framed structures of performance, the activities were framed with preparation, reflection, and return to urban life. The encounter consisted of many social performances. Throughout analysis, I began to see a less traditional kind of liminality. The liminal is a threshold, yes, a point or moment in which a person or a group could be said to have changed due to experiencing something. Yet, that threshold was unclear. The threshold seemed to be within moments of experience, conversation, and activity. Not only were these groups crossing physical liminal boundaries, but they were, through interaction, seemingly crossing various kinds of mental thresholds particular to each experience. Whether these were significant aspects of the journey or activity or small moments within, each had a transition point. There were apparent changes in thought processes seen in evident alterations in the way that people behaved. There were evident cases in which people spoke differently with regard to themselves and others. There were changes in group dynamics. This seemed to come about due to a moment shared in an ambiguous and unfamiliar setting, as simply put in the figure above.

I do not necessarily celebrate these interstitial moments[20] but merely acknowledge these intervals between states of motion, behaviour, and changing relationships. These excursions remove groups from their norms. The 'betwixt and between'[21] or the ambiguous spaces that individuals found themselves in were often intangible. The concept of liminality was used as a framework in which to place these moments and to understand what it is that happens within these excursions into natural landscapes that lead to the idea that they are in some way transformative.

> [Communitas] may be said to exist more in contrast than in active opposition to social structure, as an alternative and more "liberated" way of being socially human, a way both of being detached from social structure [...] and also of a "distanced" or "marginal" persons' being more attached to other disengaged persons [...][22]

Liminal situations manifested a feeling amongst groups that Turner referred to as 'communitas.' Communitas for Turner is the bonded, empathic, joyous essence arrived at by engaging in a joint activity and passing through a liminal threshold with others. With this may come a sense of similarities to others and heightened sociality. Turner's 'Communitas'[23] speaks of a moment of connectedness: 'An essential and generic human bond.'[24] It is 'within the liminal phase of ritual and within the liminoid of phenomena, Turner asserted that communitas could develop.'[25] For Turner,[26] liminality can also be described as a stage of reflection.[27] It is within these ambiguous moments between self, group, and the non-human landscape as well as in the reflection and sharing of these experiences that some kind of transformation, as perceived by the individual, intended by the group, and set up as the agenda of these cases study groups, occurs.

The etymological meaning of 'moment' is movement, or a changed state. Each excursion, and moment within, had the potential to be liminal and to give rise to transformation of some kind. Through meeting others, sharing experience, and entering new spaces people were changed, transforming from moment to moment with each interaction. Liminal moments of experienced 'communitas' are not solely rites of passage nor ritual excursions as in the traditional anthropological sense. Instead, I suggest that experiences and moments within these mundane excursions were liminal. Liminality was not only the outcome of these excursions, but the cause of the perception of these excursions as transformative. This has been left unsaid. In understanding this phenomenon, we understand more fully shared experience of natural landscape and the value of these excursions to human mental wellbeing.

When researching within groups, I became sensitive to these moments of liminal encounter. I saw not simply spatial, physical, or literal moments of potential liminality but emotional and metaphorical moments where the group or an individual may have been said to be crossing a threshold. This may be in relation to personal challenges, pushing personal boundaries, attempting 'inner work' or reflection, or in doing something that may have, for them, involved a departure from the norm, and a return, post-encounter, to what they considered a normal or everyday situation.

In February, on one excursion with the Young Adults in the Arrochar Alps in Argyle and Bute on the west coast of Scotland, I was caving and weaselling with the Glasgow group. We were about to take on the most challenging of caves of the day, and I volunteered alongside another female participant to go first.

The inside of this narrow and deep cave was explained to us, and we were going to do exactly as the facilitators told us. We were to wriggle our way, first down and into the cave and the darkness, and then we would come to a 'V shape,' with ledges on which we could lean our elbows on either side. We were to squeeze our hips into the 'V shape' and this would hold us up as there was supposedly a drop below. The facilitator wriggled his body to demonstrate. Samantha went in first, and she struggled as her hips were too narrow. She was unable to see in the darkness what she was supposed to be doing. She kept trying before turning back to me and asking if I wanted to come down. Getting frustrated, she asked me to go in front of her. My hips were also too narrow, and my arms were not strong enough to hold myself up. I attempted to swing forward to see what I could find with my feet—nothing. I attempted to walk forward hanging with my elbows, no luck. I tried to reach the base of the cave—'the void'—with my feet, but I couldn't feel or see the bottom. At this point, I could not understand at all how we might get across this gap. Meanwhile, Seb was at the other end of the cave (a light that we could barely see) shouting what we were meant to do. We tried to make him understand that we could not do as he said as our hips were too narrow. Samantha and I were getting a little hysterical and scared, and were laughing and trying to offer advice to one another. I turned to Samantha and asked her to hold my hand—I was going to get down and see just how deep this 'void' was—I was very tentative! The space between the wet, dank, and sharp rock faces was very narrow, and as I reached down with my feet, I could feel it getting tighter around my legs and torso and the arm rests that were above getting further out of reach. I reached the bottom, a deep stagnant water, right up to my knees and freezing cold. I was reassured at feeling the floor and decided that I would wriggle along this way. The walls began to narrow as I shimmied along with Seb saying, 'You shouldn't have gone down that far!' At the other end, I turned and dragged myself up, face against the wet, cold, scraggy rocks and sliding on my belly. I hoisted myself up and out of the cave and turned back to help Samantha up and out as she followed.

There were potential moments throughout this encounter for myself and Samantha to have felt challenged, to have felt as though we were pushing through various fears and apprehensions: to drop into a cave, to use our strength (or lack of it) to hold our bodies upright, to encourage ourselves and each other despite confusion and panic, and to trust in ourselves to drop to the floor. When we reached the other side and pulled ourselves up to be able to stand on terra firma, normality, we were filled with adrenaline. We high-fived each other, laughed, and chattered. This was liminal in two senses—physical or spatial—travelling through the rocks and emotional or metaphorical. Both Samantha and I crossed mental thresholds to move through the cave. This manifested a sense of connection, empathy, and joy after the fear. We felt closer due to sharing a mutual experience with one another. Our relationship had changed towards both the space and to each other.

When Samantha and I were contained within the cave, we were affected by the space: the primal fear of falling, being trapped, and the bodily sensations of being held up by the rocks. The emotions within the moment and the rationalisation that we came to whilst endeavouring to pass through were powerful and new. Neither Samantha nor myself had been in a similar scenario before this day. Within the moment, I was not a researcher; I was fully participating and frightened. I had no distance from the activity, but I was able to reflect on this experience, to physically, albeit subjectively *understand* the experience. By being in this moment, I felt the visceral connection to both the landscape and my informant and due to this was able to talk to her about it in a way that made sense. Without having this perspective, I would not have been able to *see* what was involved. I would not be able to speak to her about her experience in a way that was called for, more than empathic, more human.

For everyone within this group, this intensely tactile interaction was a new experience. An opportunity for the 'metropolitan body' to try something more akin to the 'climbing body.'[28] For Lewis, 'the metropolitan body' is one that is passive, inorganic, ocular, and groundless. The 'climbing body,' however, is 'organic, self-determined, and off the ground.'[29] For George Simmel, modernity has posed 'a potentially irredeemable rupture for embodied experience: [where] a potentially adventurous and corporeal body would coalesce' into the 'average sensibility' of modern life.[30] For Norbert Elias, the modern world brought with it the demise of the sense of touch and a redundancy of the hands.[31] This encounter allowed for a moment of, at least believed to be so—perhaps even subconsciously, a marginal situation.[32] For individuals within the group, and myself, this encounter was ambiguous. It was new, unfamiliar, and asked of us that we engage our minds and bodies in a way that we may not have before. We were asked to place ourselves in a situation of uncertainty and challenge. We needed to be active in our participation. There was no room here for complacency or lack of awareness of our surroundings and others. This encounter provided both physical and mental boundaries to be crossed.

The above description provides a stark image of metaphorical rebirth, the rocks representative of the birth canal. The cave was indeed a literal threshold. This was one of many symbolic thresholds within the field. When people speak of rebirth, they often refer to new life, transformation, and a sense of revival. Turner describes liminality as 'cunicular'—like being in a tunnel between the entrance and the exit.[33] These caves echo this description. Fieldwork was full of examples such as this—forests to find our way through to the vast empty spaces on the other side, rivers to cross to reach our bunk house and some form of stability, roads to travel along to reach our destination, and hills to get to the top of and back down in one piece. However, it was the metaphorically liminal that emerged as the most prominent theme in analysis. I do not use liminality merely in an abstract theoretical way but instead as a useful and practical framework stemming from observation and fieldwork—a way to structure and understand data.

Beech[34] recognises three elements of liminal practice. These are experimentation, reflection, and recognition. In my case, I amend this idea slightly. Beech's 'experimentation,' becomes engaging in these excursions and attuning to the spaces and groups. His 'reflection' becomes the sharing of experiences, conversations, and dialogues throughout excursions, and Beech's 'recognition' might be speaking of transformation felt within the self and the group. Liminality allows us to explore the possibilities for transformation—how they are spoken about and witnessed and how these encounters reflect on individuals every day. The journey and being in the natural landscape before returning is an obvious set-up but one that provides fruitful opportunity to consider what it is that happens within this structure and how each person is affected. This then allows us to consider why such excursions may be said to be transformative.

All the groups that I travelled with appeared to adopt frameworks like that of Braaten[35] and his 'Five Elements of In-group Dimensions.'[36] Though Braaten was speaking of successful group psychotherapy (of which the groups I joined may partly have affiliations), I saw synergies with this more pragmatic way of thinking. The five elements are: 'attraction and bonding,' 'support and caring,' 'listening and empathy,' 'self-disclosure and feedback,' and 'process and performance.'[37] Each of these could arguably be elements of Turner's 'communitas.' For all but one of my groups these were deliberately constructed within the agenda of the group. For The Woodland Weekend Group, this was not explicit though within The Mental Health Initiative they were. You could say that each group actively encouraged communitas, perhaps by a different name. The Young Adults group adopted some of the earlier group dimensions such as 'resolving conflict and rebellion,' 'constructive norming and culture building,' and 'reducing avoidance and defensiveness'[38] also a deliberate set up of personal development activities.

Liminality emerged as an outcome in these encounters in the field. This was in relation to the perception of one's self, the group dynamics—intentions, obligations, and communitas—and subsequently how then the landscape may be a liminal site not only physically but metaphorically in terms of the way it is perceived and experienced. This is to do with how a space becomes a place of meaning and how the self, human, and landscape interrelate in making this so.

The Sense of Self and Group Dynamics as Sites of Liminality

> if you continually do that and you continually, 'I'm not going to engage and I'm not…', like you just box yourself into your bedroom, as some of them have and then start collapsing as a person because we're human beings we need connections we need to be pushed you know we're part of, our kind of evolution is about adapting to our environment and I think a lot of our skill sets are about that and you need to kind of, part of being alive is that you need to keep pushing yourself and expanding yourself and trying different things and trying different sides of yourself. Because that's the human being, I think that's the human condition and you need to do that because that's the way we've developed […] What if you don't do the big jump you know? and that means you're not engaged in that situation which is fine for now. But what if you continually don't do that big jump? […] (Rory)

On one occasion, the support workers and I sat in a board room at the base of The Young Adults' Personal Development Project. We had been talking about anxiety—specifically, the fear of doing something we have never done before. We drank coffee. The 'young people' were not in that afternoon. They took one day off per week and hence the offices were quiet. There was no screaming and shouting, no heads were popping in and out through the doors. It was still, silent. Rory and Anna were discussing with me the need to engage with activities that push us out of our comfort zones. They discussed the need to do things that we are unsure of with other people—allowing ourselves to make new, human connections. For Rory, it is a 'part of human evolution to allow ourselves to be challenged and changed.' To leave oneself stagnant and unchanged is for Rory then, inhumane. One wonders whether this may be applied to a necessity for rites of passage, at the very least, moments that are liminoid—momentary actions that have the potential to change us from within.

I consider here the sense of self as a specifically liminal site. Beginning with the concept of 'suspended identity,'[39] an identity in flux I consider the figurative gap between the actual self and the ideal or ought self. This misalignment became a threshold which, throughout activities and engagement, individuals metaphorically aimed to cross over.

We have seen in the previous chapters that there are two main motivations for engagement: to self-verify and to feel a sense of belonging within the group. Within the sense of self, there is a transformation due to a metaphorical shift in perception of oneself within the group. This, I argue, is a *metaphorical liminal threshold*. This has laid a suitable path allowing me to go on and explore the group of individuals, collectively as a transient liminal site. By first exploring group development theory, more specifically Tuckman's[40] model, I then draw similarities with the stages of liminal encounter as outlined by Victor Turner to discuss how these activities allow for new-found group cohesiveness and emerging 'communitas.' This then opens out a conversation regarding 'collective intention,'[41] 'we-intentions,'[42] and 'collective obligation'[43] amongst the group. I then extend these ideas to the perception of the landscape.

I consider firstly the perception of the landscape within both the self and the collective as a liminoid phenomenon and secondly how, through building context and experience of a space and in relating the self to the landscape, perceptions of the space and opportunities within allow them to become places of meaning to people. This chapter sets out to identify these three sites—the self, the group, and the way that the landscape is perceived, as liminal. It will set out how, in their relations, these three liminal sites allow for positive personal and social transformation.

The Metaphor

I insist on both 'liminal' and 'threshold' to refer to this metaphor regardless of the seeming tautology between terms. 'Liminal' refers to the transience of the experience—the sense of temporality and ambiguity with which the self is placed. The sense of self is liminal in the sense that it is in flux throughout these experiences. The ambiguity and temporality of these activities is key in this sense of liminality. This is considered in relation to placing one's self out with the everyday, in an uncertain and unfamiliar scenario for a temporary period before returning. I look at the effect of this on perceptions of the self. I then consider what it is to shift from being a person with individual agency to part of a collective with 'plural subject[hood].'[44] Ultimately, I explore what it was for members of the groups to feel that sense of belonging that was the motivator for taking part. The 'threshold' within my phrase is of course not a physical one. The 'threshold' is where the metaphor lies. It is important to understand that there can be no literal threshold in a mind that we cannot strictly locate. The threshold metaphor helps us to understand these kinds of personal transformations (Fig. 5.2).

These excursions entail both embodiment and cerebral experience within particular landscapes—a pairing that could usually be said to aid in the development of the sense of self. Here though the experience and landscape could be said to transform due to this particular set of variables. The excursions are a process of lots of transactions between the mind, body, environment, and group, with the latter three making up the extended mind. The figure above displays how one might envisage this metaphor. The metaphorical liminal threshold (if one was to assume that the mind is only within the skull) represents a moment of transition between the self-pre-experience and the self-post-experience. This is influenced by everything within the extended mind and within the

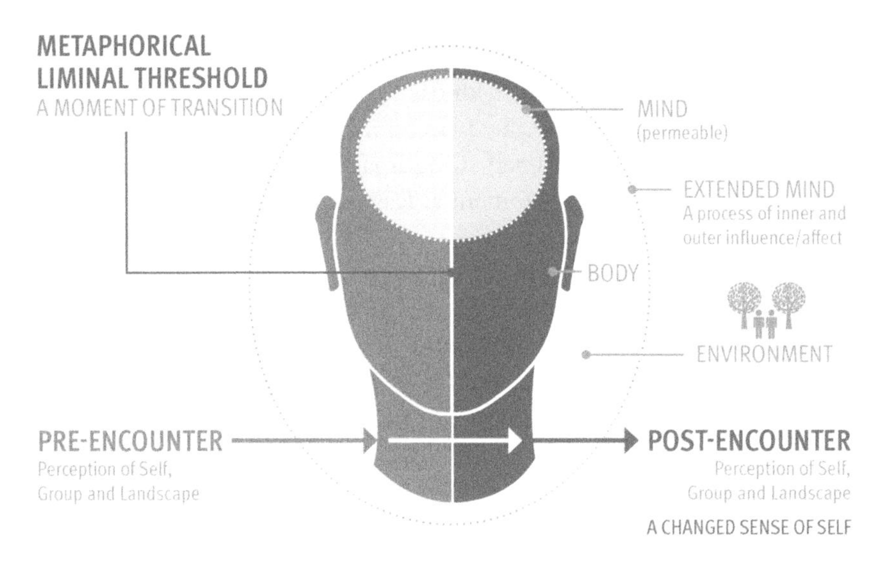

Fig. 5.2 Liminal threshold in the mind/the self as a liminal site—a metaphor

scenario. The metaphorical threshold is in the mind between the actual self and the ideal sense of self—encounters allow a crossing of this threshold and way of thinking of oneself.

> I think that there's something that happens where if you're in nature for long enough you automatically feel more yourself and you feel stronger. Now whether that comes from you or that's the trees talking to you [laughter] or whatever it is that maybe has to do with people's belief systems I feel like this is potentially like a spiritual question but I feel like it's almost, like nature allows for automatic and natural transformations to take place (Holly)

Here, we have that intangible 'something' once more; whilst individuals seek something, they also are aware that something happens. *Something* undefinable occurs when people interact with natural spaces. For Holly, she equates this to feeling more herself, to feeling stronger. If we consider that Holly is a member of The Mental Health Initiative, we might assume that, when she suggests she feels stronger, this is a mental state of being abler to deal with life's situations. Being in natural space alters her state of mind and allows her closer to a positive ideal mentality. Holly specifically identifies that for her, 'nature allows for automatic and natural [...] transformations.' This space and these kinds of activities allow her sense of self to be verified. She, when in these situations, feels her sense of self fall more in line with her ideal self. One imagines this is not necessarily immediate; there must be a moment at least between feeling the way one did prior to and the way one does post-engagement. She also jokes about whether this is something that comes from within or whether this is something extrinsic, an outside influence altering the way that she feels. We will look at the idea of the non-human as agent within the following part but for now will concentrate on the changing mind.

Noble and Walker[45] speak of a 'suspended identity' within liminal moments. Entering into the unknown territory of new groups, new locations, new scenarios, learning opportunities, and often head spaces that have been, in some cases *shut down* or left unexplored leaves the actual self in some form of limbo between the before, the now, and the after experience. These locations and these shared experiences allow for personal transformation. How then is this the case? I argue that it is in the self metaphorically containing a liminal threshold that is *crossed* throughout experience. In gaining context, new insights, self-verification, place identity, and a sense of being a part of a group, the threshold between actual and ideal self is crossed, perhaps temporarily but nonetheless useful in understanding perceived positive benefits. This temporality further likens these experiences to those of van Gennep's liminality. For Turner,

> a rite of passage is less about shifting social status and more about personal transformation-a process of transition from which one returns to society empowered by renewed creative energy, an expanded worldview, and a greater sense of hope.[46]

Though I would not say that each excursion is necessarily a rite of passage, Turner's assertion that a liminoid moment involves a personal transformation is apt. Particularly in that he believes that individuals will return, having spent time outside of their everyday situation with a new way of perceiving the world. Turner's 'betwixt and between' is pertinent here; for me this is perhaps the limbo of which I speak. This limbo separates the actual self and the opportunity for the ideal self to be verified within these excursions. According to Meethan,[47] Turner's original model—separation, liminality, and reintegration—has since had its scope widened to involve 'more general states of ambiguity.' I believe that the ambiguity found preceding and during encounter, and the temporality of these encounters, is what forces the self into challenging and transformational states. For Thomassen,[48] 'the temporal dimension of liminality can relate to [...] moments [...] periods [...and] epochs' allowing scope for these encounters, whether day excursions or six-month long courses, to have the potential to be liminal.

Ross describes a scenario within The Mental Health Initiative which poses several questions:

> one person who used to be sick from anxiety, you know turning up at this site being sick from anxiety and not able to get on the bus because of that, and then after a few months, he didn't really speak to anybody, and after a few months [...] He'd be kind of joking with people he felt comfortable

> with and that was a huge transformation [...] it's still progress to make it happen and the other thing with that person was the lowering of anxiety took a long time to happen but the kind of purpose when that person was given a job to do and threw himself into the physicality of it, was instant you know. He went from just not knowing what to do with himself to just being kind of alive in a way. The same way that I felt when I went from you know working in a hospital setting to working in the woods and just feeling like it felt right. (Ross)

This speaks of a man who was crippled with anxiety. His transformation was not so much about attaining the ideal self but instead a transformation that allowed him to be able to interact with others in a way that he had not done before. This man's anxiety was visceral, intense enough to make him physically sick. Though Ross states that it took some time, he does not state the length of time; however, knowing what I know about the project, I would surmise it being less than one year. In this time, the man changed. His anxiety was lowered, and he appeared to feel safe. He was able to joke with others within the group. One gets a sense that much like my informant group, attending the same project, this man began to feel like he was a part of the group. He began to associate himself with this group and with this space, leading to a sense of safety, or perhaps even that ever desirable belonging. This was a shift from fear to comfort within social situations. Ross describes this shift as 'being kind of alive in a way' and likens what he saw within this man's behaviour and character to a time when he chose to leave the National Health Service to 'work in the woods.' Ross made this decision due to not feeling himself within the confines of institutional medicine and working indoors. For Ross, this move was to allow him to work in a way that better suited his perception of himself. The man about whom he spoke however came to the initiative to feel *better*; without knowing this man it would be impossible for me to say that he too sought self-verification. I take Ross' word that he felt *better*, he behaved in a more animated way, he was no longer sick with anxiety—*something* changed within this man's sense of self.

> Carnival reversal implies a change from principles of stability and closure to constant possibility [...] laughter and excess push aside the seriousness

> and the hierarchies of "official" life [...] Carnival shakes up the authoritative version of language and values, making room for a multiplicity of voices and meanings.[49]

Let me take this in a seemingly obscure trajectory for a moment. Within all of my case study group activities, there is a subversion of the norm, some drastic and some subtle—normally dependent on the contexts and experience of the group. If we consider a Bakhtinian subversion of norms,[50] an allowance for a different kind of life, an unofficial life, led by participants, a role reversal in which the participants form the structure of days in part, this is where by normative standards *play* becomes *personal work*. This is where the collective is given licence to do something different and in doing so rethink matters pertinent to themselves. In these scenarios, people may give a different social performance to what they may usually. I believe Bakhtin's principles may prove useful when considering the success of these excursions. Within The Mental Health Initiative, volunteers were working in new mediums with flexible hours, no walls, and new social contracts and protocols, certainly out of the ordinary for those who had worked a full-time office job, at college, or in a cafe. The Woodland Weekend Group left the urban behind in the hope of complete subversion: tent camping, sleeping on village hall floors, and physical and *get your hands dirty* tasks with a new social circle. For The Loose Community, the subversions include staying somewhere new and allowing for an attempt to influence one's own ways of thinking through attending, with an open mind to the landscape, and in being willing to share in new activities. For the Young Adults attending the personal development project and The Community Living Initiative, these subversions seem much more drastic: no electricity, no phones, no contact with home, new people, new leaders, new hierarchies, new rules, new social contracts, and new challenges.

One subverted feature of these excursions in respect of what is deemed normal to the groups in their everyday is the structure of these excursions or lack thereof in some cases. Another distinctly liminal feature is the concept of anti-structure.[51] How might structure effect someone's perception of themselves? How might change in structure allow for personal transformation?

> I was in a state where I couldn't cope with structure but I couldn't cope without it so I felt really, really strapped, trapped so the fact that this was enough structure that I could deal with but not too much enforced structure then... [gestures—was positive]. (Jess)

> human beings aren't very good without structure, they really aren't, well I certainly amen't [sic.], you know, you disintegrate as a person if you don't you know? and normally you do school, if you go to college or you start work, your structures taken care of and you don't think about it do you? (Greig)

For individuals within The Mental Health Initiative, it was in fact a need for structure that proved important in choosing to attend. Greig was no longer working, having been signed off from his office job due to mental health 'problems.' He felt that he had no stability because he did not have any daily or weekly structure. Jess described her situation before she joined this project as being in a state of desperation with a need for some kind of framework for her life but unable to cope with enforced structure. She had previously had difficulty attending work and college. Interestingly here, The Mental Health Initiative provides a very loose structure—a morning arrival time, lunch time, and an end time. On days where we were at the peri-urban base this was the extent to which the day was formally structured. Within this time though there was flexible time to work on something, a selection of tasks were outlined at the start of the day, and individuals were asked if they would like to contribute to each task. A structure remains, but this is a framework of new structures. Jess was unable to adhere to normative modern structures associated with urban life and so struggled. Within this initiative though she found a new kind of structure that allowed for her to be more comfortable. This was a similar sentiment to Holly's:

> I think that's a super important point like, and that's the thing that's missing for a lot of people and that's you know it's so crazy isn't it because it's like the structure, the normal structures that exist are two oppressive or stressful and then no structure at all it just makes you turn in on yourself and be self-destructive as well so it's kind of like whoa! at a certain point it feels like its lose, lose and then how do you bring yourself out of that? I

> think if there were more projects like this, maybe each one slightly different it would just be amazing so, you know, for me personally I'm just lucky enough to have found this that happens to suit me, like happens to suit the balance that I need.

For Greig, structure is something that needs to be provided. In our early years, it is provided by institutions of which we must become part (e.g. school). As an adult, however, one must think about where these structures may come from, and for Holly and Jess the normative structures that one may find readily available were untenable (e.g. a 9 to 5 job). It seems then that structure or a need for a different manifestation of structure was needed by Greig, Jess, and Holly. Holly saw this need in many people. She finds 'normal' structures, those found in schools, colleges, and work places as oppressive. They do not allow her to be herself and thus she struggles to conform, and she struggles to survive. These structures increase her levels of anxiety to the point of seeking *something* else. No structure at all, echoing Greig's sentiments where he felt one would 'disintegrate as a person,' led her to become 'self-destructive.' A lack of structure leaves her to flounder. The Mental Health Initiative provides a balanced framework in which there are rules, there are expectations, but these are more suited beyond the normative. Providing this flexibility, allows both Greig, Jess, and Holly to feel that they may thrive.

Within The Young Adults' Personal Development Project, a key feature was providing a sense of structure to the young adults who are considered to have disrupted and unstable outside lives. As a project, that intends to 'prepare for work,' this is not surprising. Outside the remits of arriving on time, administration, health, safety, and agreed acceptable behaviours (and repercussions) within this project there is no labour, but exploration, freedom from adult constraints, space to be idle, spaces that allow for new behaviours, and changes in role performance. John, the project manager, assured me that this course was intended to be led by the 'young people' allowing them the opportunity to make decisions, an opportunity possibly seldom afforded to them in previous similar circumstances. I see this in the later parts of the course, when the group members are asked to plan their own walking route, to decide where they will camp and when, and to take the lead in managing themselves as a group with the support of the leaders standing by. It is evident in the behaviours of the group facilitators, bringing themselves down to at least a linguistically and aesthetically equal level. In language, they assert an equality of roles, if not quite in action. One participant from this project considers this differently. Keith has a significant amount of routine in his life, and he values it. Keith instead is learning within himself to be comfortable with the changing of routine and regularity, a different personal challenge:

> I do value routine yeah, but sometimes you've got to let the routine go and get outside the routine cos sometimes not everything goes according to plan. So it could be that one thing in the routine half way through just slips up and then you have to sort of change it and adapt and do a different routine (Keith)

It strikes me that Keith may in fact be referring to not only routine but ways of thinking and reacting; he stresses letting go of the routine and getting outside of the routine, and he speaks of adapting and doing 'a different routine.' In this conversation, he was comparing himself to others in the group and suggesting that he may be more capable of adapting to challenge and change than the others. I believe, due to knowing his character and vocabulary that when he says 'routine,' he means planning and his reference to flexibility refers to being not only able to plan but to adapt in new-found situations. Keith considers himself to have more experience than others, he is the only participant with significant experience outdoors in his work with the Cadets and the Scouts as a child. He feels familiar with rules and structures, and he appreciates them. For Keith, then, he welcomes the subversion of the predictable as a challenge, an opportunity for him to think outside the box.

In idle and unstructured time, there are opportunities for social interaction, for new relations to flourish, for conversation, for reverie, reflection, and solitary thought. In an environment where idleness, challenge, and potential contributions to a sense of being in flow are actively encouraged as a means to becoming more productive and content as an individual, new awareness emerges. In new awareness we find, perhaps, a new sense of self. When removed from normative structures, adult responsibility, and constraints and instead asked to contribute to collective aims and to share collective priorities, modes of thought change. This may only become evident on finding that sense of belonging when the liminal threshold between actual and ideal self has been crossed. A sense that one belongs allows people to begin to consider themselves as part of the collective *we*—they are no longer an individual, still ambiguous of their place within the group but instead a member of a clear, cohesive collective.

> [...] everybody's expected to attend, go to reflections in the morning or reflections in the evening. (Allan)

Reflection allows an opportunity to consider experience in relation to the self. Group reflection however allows an individual to openly discuss these experiences. Earlier, I referred to psychoanalyst Wilfred Bion's group reverie, considered as an activity of transformation,[52] where the contact barrier in the psyche becomes elasticated and where the sense impressions related to emotional experience are transformed.[53] Bion believed this to be possible across two people—reverie with another or 'the work of construction and searching for significance and attributions of meaning.'[54] For me, this holds possibilities for truly imaginative empathic engagement, valuable dialogue, and perhaps even a collaborative knowledge that helps in the changing dynamics of the group. In his discussion of 'the container and contained,' Bion[55] speaks of the potentials of empathy and reverie with another. He states that

> at critical moments, to develop thoughts realistically, the human self needs the social container of pair and group to actively participate in meaning making

Group reflection, as an activity found across all case study groups (sometimes by a different name i.e. sharing, discussion, debrief, or feedback) furthers similarity between case study set-up and Turner's liminoid. It is an activity which explicitly invites participants to communicate their experience in the now. For Turner, 'liminality is the realm of primitive hypothesis, where there is a certain freedom to juggle with the factors of existence,'[56] and therefore 'liminality may be partly described as a stage of reflection.'[57] Reflection time, story sharing, or time to provide feedback at the end of an experience is common, and with each case study, there is an emphasis on a deliberate lack of hierarchy, mentioned by Allan when he says that no distinction is made between staff and group. Whether this activity could be considered as mundane or sacred or indeed therapeutic—discussion allowing for reflection on experience is encouraged. For Beech,[58] reflection is a practice that allows for an emphasis on 'both outside-in and internalised dialogue'; within this, there is 'self-questioning,' 'self-change,' and a reaction to or 'absorbing of external influences and perceptions.'[59] For Cuncliffe,[60] this kind of active questioning of how we are perceived internally and by others is 'the essence of a dialogic construction of the self.'[61]

Reflection, conversation, and sharing are key to these experiences of the natural landscape because this kind of social interaction allows individuals to consider their experience, themselves within these contexts and as a member of the group, and in doing so reconstruct or reimagine what their self is. The individual, on reimagining himself within a group, may move towards a better understanding of the self that is perhaps more closely aligned with positive conceptions of the self. They also bond with others within the same practice. A consideration of both the self and other in context allows for empathic behaviour and a shared set of priorities. Within these new structures, self and group reflections, and new developing sense of identity, a metaphorical threshold in the mind is figuratively crossed. Within carrying out these activities with others, this threshold becomes a site of transformation (Fig. 5.3).

The Group of Individuals Collectively as a Liminal Site

The developments within the groups are key to understanding how these events may be considered as transformative. Whilst above, I have described the self as a liminal site this of course is in part to do with the self's interaction with the group that they are found within. This I would

Fig. 5.3 Some members of The Community Living Initiative group take a walk together on the beach

argue feeds into the ever-developing group. The self is altering and thus altering the dynamics of the group with each person similarly going through these ever-emerging liminal stages and contributing to this. The group is going through a liminal transition from a group of individuals to becoming a collective, a 'plural subject'[62] perhaps in the manner outlined by Gilbert:

> [O]f a number of persons (two or more) has, in effect, offered his will to be part of a pool of wills which is dedicated, as one, to that goal [...] a binding together of a set of individual wills so as to constitute a single, 'plural will' dedicated to a particular goal. [...] refer to themselves as 'us' or 'we', to 'our goal' 'our belief ', and so on...[63]

When people consider themselves within these situations they consider themselves within the group. The individual, whilst still of course an active and unique agent, is no longer a person alone entering a group but is now part of a group of actors. They have had the intention to be part of the group and willingly share in activities and thus experience. People consider their role within the group and they witness how, over time the group dynamics changes. This shift in dynamics inevitably repositions each individual and puts one's sense of self in flux, in relation to the behaviours, actions, and feelings of others. The changing group dynamics changes experience. I additionally suggest that this dynamics aids in the process of self-transformation, with positive consequences for individuals. The sense of belonging discussed throughout becomes a sense of communitas, brought about due to sharing in experience, witnessing to changing dynamics and the subsequent development of the self. The group begins to form friendships, shared contexts, memories, collective intentions, and sometimes a sense of collective obligation. This then has significant impact on how the overall scenarios are perceived, and personal and group aims and agendas are altered. Ultimately, this has considerable impact on the opportunities seen within the landscape, which again is ever changing. The development of the group is a process, with visible stages.

> So you often see people nervous or anxious or, maybe sometimes that comes in sort of value defensive behaviour, not wanting to do things or kicking off a bit, and so typically the first couple of days, the first day that they're here it's all a bit, 'what's going on?' and then the second day it's like 'I'm finding this quite challenging' and then by, by the end of the third day they start to settle into it and start feeling more confident and are able to explore and give things another go and enjoying their friends and then realising, you know, the amount of freedom and safety actually in comparison to what they might have back home where, you know, the threats are quite often real. (Allan)

The individuals that travel to The Community Living Initiative first make the journey across Scotland before arriving either as part of a group or on their own. For these people, both the unfamiliar surroundings and the group in this context are new. Joining a group is of course the initial stage in having an impact on a group dynamics. For those that have travelled here, they are now separated from not only their urban environments but also people in their everyday life. For those beginning at The Young Adults' Personal Development Project, the initial meeting is at the base before they travel anywhere. For those in The Mental Health Initiative, as we have seen, this initial stage is upon walking through the gates of the peri-urban site. The Woodland Weekend Group meets first at the bus collection point, and finally, The Loose Community meets at various locations out of the urban area. These individuals, whilst Helen, Jill, and Catherine may be familiar with one another, are often invited by one member of the group and meet the rest for the first time on site. All individuals though, at one point, were independent people who now are submersing themselves within a new group in an unfamiliar place (Fig. 5.4).

> Forming [...] In this stage, the group becomes oriented to the task, creates ground rules, and tests the boundaries for interpersonal and task behaviours. This is also the stage in which group members establish relationships with leaders, organisational standards, and each other.[64]

If we consider Allan's comment regarding visitors to the island we see again, as with The Mental Health Initiative, that the experience begins with anxiety and uncertainty. They wonder, 'what's going on?' In Allan's experience, this comes along with the exhibition of value defensive behaviour. The defence of one's values could be considered as a mechanism of fear, as again, we have seen before, of being unable to fit in and unable to self-verify. This is potentially a fear of the unknown. This, in much the same way as within The Young Adults' Personal Development Project seems to bring about a refusal to take part or a reluctance 'to make that jump' as Rory described. This is an apprehension that can be seen across all of my case study groups. What Allan describes could be considered an example of 'storming.'[65] Tuckman's 'storming' stage, following the 'forming' stage, is considered as the stage in which the group grapples with 'intergroup conflict.' It is a time characterised by a 'lack of unity and polarization around interpersonal issues,' where people 'resist moving

Fig. 5.4 The Loose Community members walk together in Callander

into unknown areas of interpersonal relations and seek to retain security.'[66] The individuals, according to Allan, 'kick off a bit.' This is colloquial, referring to a sudden anger or upset that may lead to fighting or arguments. This is an accurate, if informal, description of what Tuckman was speaking about.

> 'group members become hostile toward one another and toward a therapist or trainer as a means of expressing their individuality and resisting the formation of group structure'. In this stage, members may have an emotional response to the task, especially when goals are associated with self-

understanding and self-change. Emotional responses may be less visible in groups working toward impersonal and intellectual tasks, but resistance may still be present.[67]

When Allan's participants exclaim 'I'm finding this […] challenging,' he suggests an openness to articulate difficulties in a new, more positive and clearly communicative manner, perhaps referring to a shift towards what Tuckman[68] would call, the 'norming' stage. The 'norming stage' is where the group begins to 'develop cohesion' and 'express personal opinion.'[69] It is within this stage that roles and norms are established.[70] Allan's group begins to settle into the community, and indeed the unfamiliar environment, and to feel more confident. There is an understanding that there is no tangible threat, a common suspicion amongst groups, particularly the Young Adults but which could also be seen in the feelings and behaviours of The Mental Health Initiative members. The group form friendships and they feel a freedom to simply be there and to safely explore, comfortable with their idleness and island exploration. Finally, when groups reach the peak of Tuckman's 'norming' stage, the group becomes an entity as members develop in-group feeling and seek to maintain and perpetuate the group.[71]

Allan's anecdote provides a concise and typical explanation. Perhaps not always so straightforward an example but I feel a truthful representation if not a metaphor for all initial group encounters. This is, from what I have witnessed, a very familiar scenario. Sometimes more or less volatile than this however with similar outcomes, confidence to be open and exploratory, a sense of place, and a discovery of how one fits within a group. This is the case not only between the participants but in their relations towards the facilitators too.

Storming and Norming

As we prepared to leave for Bonaly, Jonathan had not arrived. He was coming from Harrogate and had slept in again. It was quite standard to wait for a considerable length of time for everyone to arrive before we headed out, but on this occasion, we were more pressed for time. As it was coming into March, and the final 'bothy challenges' were looming, the facilitators felt a more pressing importance to get out and spend the most time possible outdoors on the day trips. The other participants were also becoming fed up with waiting for others.

Usually, it was believed to be imperative that each individual packed their own backpack with all of the kit that they may need. It was also considered important that every member prepared their own lunch. This was part of the drive towards teaching the skills necessary to look after oneself. On this day, Rory decided to pack Jonathan's bag to speed up the process, and Paul carried his kit to the bus and at the beginning of the walk at Bonaly. Anna made his packed lunch, from a selection of rolls, wraps, hams, and salad laid out in the back room of the base.

When lunchtime arrived, we sat at a crossroads on the hill, one path was leading all the way up, one was leading back down, and one was crossing the plateau. We were scattered around. Everyone pulled their see-through sandwich bag, squashed bread rolls, and crisps out of their backpacks. Anna had made a roll for Jonathan. He pulled it apart tossing the plastic bag and salad to the floor. This was a deliberately defiant action in which he willingly ignored an explicit group value, to conserve the landscape. It was also aggressively flippant with regard to the efforts of other group members in carrying his bag, packing his kit, and making him lunch. When this was then met with the request to rectify the situation, Jonathan refused point blank, purposely and clumsily kicking the flimsy bag, which barely moved on the sodden, muddy ground:

> F**k this shit, I'm not picking up the f**king salad, it's your f**king fault for putting f**king salad in it!

Anna asked him not to speak to her in such a way but was almost immediately interrupted by Amy:

> You want me to batter him for you Anna?

This was said in a tongue-in-cheek manner. It was an insight into Amy's developing relationship with Anna and her perception of Jonathan within the group. It was said in a way as if to imply 'I've got your back

Anna.' It is met with a laugh from both Anna, regardless of the tense situation, and the rest of the group. This added to Jonathan's frustration. I believe by this remark Amy was intending to communicate a feeling of mutual respect with the group leader. The reader may wonder how on earth proclaiming to 'batter' someone could have this effect. This is a common remark amongst the group, particularly between Amy, Sam, and Jonathan. It is normally said if it is felt that someone is being picked on, regardless of whether in a serious or frivolous manner. By doing this, I believe Amy was not only trying to show a little camaraderie with Anna but also to defuse the situation.[72] Witnessing Amy's tone would suggest that this was not meant to be taken seriously, perhaps suggesting that she did not take Jonathan's behaviour seriously. Clearly Amy knew that Anna would not agree with this offer; however, it would affect the situation by inadvertently showing Jonathan, due to Amy's popularity within the group, that his behaviour was not condoned by his peers.

Though this may appear to be an instance of 'storming,' there are elements within this scenario that would suggest that the group is beginning to, as Bonebright[73] may have anticipated, 'develop in-group feeling and [to] seek to maintain and perpetuate the group.'[74] Perhaps Jonathan was having a difficult day, or perhaps he was still struggling with the group dynamics and his own sense of self within the group. This may suggest that Jonathan was behind the rest of the group in terms of development; however I'm not sure that this would be the whole truth.

Each day with this group tended to throw up evidence of individuals not having yet developed this in-group feeling. It may appear that Amy had moved on; however, I'm not sure that this is the case either. Perhaps Tuckman's linear and progressive model doesn't quite fit this group's dynamics? Amy also had several moments that would put Tuckman's theory into question. These moments also further ideas regarding one's perception of the self not bearing resemblance to the actual self, particularly in behaviours. You will remember Amy's comments regarding Jonathan's maturity, and it is interesting to consider these now. The following scenarios involving Amy were later within the course.

> This place makes me angry, I think about coming here and it makes me pure raging [...] I just want to smack them across the face with a f**king frying pan, when they open their mouth they make me angry. (Amy)

Is this another example of being unable to self-verify with the group? The group evidently frustrated Amy; however when asked there was very little she chose to divulge. In April, I had a conversation with Rory regarding Amy. He described her as 'going nuclear,' inferring her rage. To prevent this, she was given consequences including three chances before she was to be taken home and potentially asked to leave the course. He had been concerned after the Young Adults' first trip to the bothy.

At the bothy, an incident arose after Amy was not given paracetamol by the leaders on her request. Amy admitted to me that she regularly took paracetamol as a means to combat stress. This resulted in a group coup against the leaders, where Rory struggled to regain the support and understanding of the participants. He was essentially 'ganged up on' after Amy had suffered from a panic attack, or at least threatened that she may have one. Against his better judgement and in his own words to 'calm the situation down,' he gave Amy the drug: 'She had won, she was in control' (Rory).

This is another example that may point towards a storming group. However, there is still strong evidence here for an 'in-group feeling' and a desire to perpetuate the group. Of course, this is not the entirety of the group as the actions clearly force the leaders into an *out-group* situation. Perhaps Amy had been left feeling vulnerable due to not being allowed her usual vice or because this had been the first time that she was far from her child. Whilst there was evidence of a changing dynamics, and perhaps of the storming qualities of which Tuckman outlines, there was more to the situation. It certainly does not seem like a model such as Tuckman's is wholly applicable.

Behaviours within The Young Adults' Personal Development Project continually fluctuated. The group development process was not linear and suggested that the self in flux was a continuous force upon the group dynamics. It may not have always been linear, but each circumstance provided scope for group changes. This is not to say that the group never reached a version of 'norming' or of Tuckman's 'performing.'[75] Instead that each moment was a new and integral ingredient impacting the dynamics and showing that each individual self, with their own individual emotions, uniquely affected this. A group is, instead of being formed, perhaps *tangled* in this instance. By this I mean it is entwined, messy, often chaotic. The group may reach a kind of stability or 'norming' in some sense but also has the potential for dissonance. Each individual adds to this dynamics.

Rory described the project as having to form a new and individual relationship with each 'young person.' He insisted that not only do they push the group leader's boundaries but that the group leaders also 'push back.' For Rory, this is not unlike any other relationship in his own life. What he perceives as differing within this group and with what he has to provide, is a stable, continuous, fair, and always caring relation. This is not a kind of relationship that he believes the young adults are used to in their everyday personal life. Instead they suffer in unstable relationships and because of this they have 'outlandish coping skills [...] like just stamping and shouting to get what they want.' He believes this is because this is what they have had to do in the past to get what they needed. To combat this the leaders had to set up firm boundaries, as they did with Amy to which she 'responded massively' because 'maybe that's what she needed.' Rory acknowledged here that he did struggle with Amy and joked that he wished that he had figured this out for her at the start of the course. He also acknowledged that it was a learning process, despite their experience, for the leaders too.

Disruptive group dynamics are not unique to the personal development projects where the participants were dealing with behavioural issues. Within my case study work with The Young Adults it was perhaps more evident that there were group development issues throughout. The group was often volatile, arguments were frequent, and the general atmosphere was perhaps more uninhibited. Perhaps other groups could be considered as more readily cohesive, such as The Woodland Weekend Group, The Loose Community, and The Mental Health Initiative. They had pre-established similar ideals. These were evident in the fact that they had joined a group specifically aimed at socialising and sustainable endeavour, though with their own individual motivations, these motivations did tend towards the less pragmatic. Regardless of shared interests and a display of pro-social behaviour they too sometimes suffered in the initial group development stages, and like The Young Adults, this may too arise throughout experiences, regardless of the amount of time spent together. If we take a scenario with The Woodland Weekenders as an example:

> I've seen two disagreements over all the years I've been and that's all which is quite a good record considering all these people, these strange people have been thrown together and one of them was a storm in a teacup [...] one of them was in a bit of a grumpy mood that day anyway so it was only a matter of time before he fell out with somebody (George)

Though George states that he had only witnessed two disagreements over his nine years attending the woodland weekends, he described one incident that was resolved 'almost straight away' and another, that he considered major. It resulted in 'various bits of paperwork flying about' as well as official complaints to the organisation. Whilst volatile situations are common amongst The Young Adults and dealt with in a very different, more immediate manner, a disagreement of this kind within The Woodland Weekend Group had, previously to George explaining to me, been unthinkable.

> it was quite a shock, I remember talking to [Rosa] about it and she was physically shocked by it, like people having an argument, two adults having an argument over a bit of wood or something. She was quite shocked about it. It affected her weekend because she actually said that she was considering not coming back to a weekend because of it because you know the atmosphere is very relaxed and everybody gets very laid back don't they and to see somebody having an argument... (George)

The incident that George described had, according to his own account, been likely to happen as the two individuals had had altercations on previous weekends. Nonetheless, it was out of character for the group as a whole. A dynamic shift such as this, though rare, I believe still has the same potential as Tuckman may say that storming has for positive group development. George believes that the reason situations such as this are so few is due to the space that the group has when out in the forest—allowing people to work separately from the group and to avoid confrontation.

For Braaten,[76] Tuckman's model is simply about becoming cohesive. Whilst his model may not be appropriate for all of my case study groups, I see that confrontation, ambiguity, challenging relationships, and navigation of the self within the group, sometimes with volatile outcomes, is an integral process in the positive transformation of groups. Though limitations within the model are evident, it is a helpful tool in beginning to understand group developmental dynamics and allows correlations between frameworks of liminal encounter to be examined in relation to the merging of individuals and in the becoming of a group.

Braaten proposes, as an alternative to Tuckman's, a five-factor model of group cohesion, that specifies the necessary components of becoming cohesive as a group:

> These factors are: attraction and bonding, support and caring, listening and empathy, self-disclosure and feedback, and process performance and goal attainment.[77]

Throughout my own analysis, I too have developed a framework that is applicable to what I have seen in the field. Having considered Tuckman's model, Braaten's five factors, and Turner's framework of separation, liminality, and reintegration I suggest an alternative. For groups to form positive and beneficial relations there seem to be three factors: (1) a desire for belonging and self-verification, established through (2) shared experience culminating in (3) a sense of collectivity or '*communitas.*'[78] This collective feeling accompanies discovered self-verification, a sense of belonging, and a sense of place. Post these three stages are where we may then see the liminal self and group. In the figure below, I show how these concepts may be considered alongside each other (Fig. 5.5).

In the evolution of groups, each has a beginning, a joining, and a separation from the group that we know. Each has a navigation period and an ambiguous and unpredictable submersion into the group encountered in sharing new experiences. This is where roles emerge, as the self-transforms and as the group becomes cohesive. This I will liken to Turner's liminoid. Each has an outcome similar to what Turner terms 'communitas,' a sense of cohesiveness, bonding, and support. In this instance, the group re-establishes norms and becomes responsive to one another as a newly established group. As the group renavigates its dynamics and each individual affirms their place within that social group, the self continues to adapt and transform.

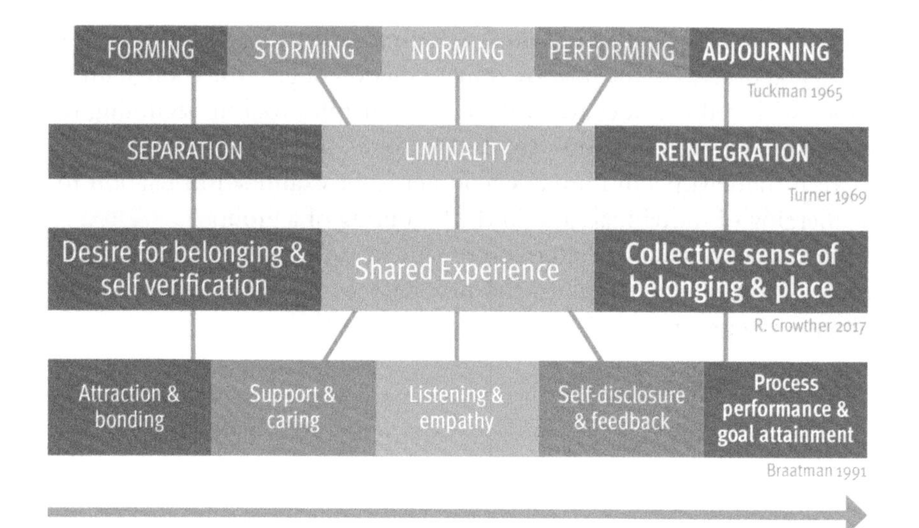

Fig. 5.5 Stages and phases

> Senses, intellect, and emotions join together in a gestalt. The cohesive group can help you become more the person you always wanted to be away from rigidity, feeling defensive, vegetating instead of living a full life in the here-and-now.[79]

This sentiment from Braaten echoes many individuals within my case study groups as I have shown. By joining a group that in time became cohesive, informants felt more in line with the person that they would like to be or believed they could be. Turner states that in 'communitas' there is no merging of identities; instead identities are liberated from normative conformity creating relationships that do not 'submerge' one identity in another but instead protect the groups' multiple differences through realising their similarities in communitas.[80] Whilst a group becomes cohesive and begins to develop an identity throughout potential liminal moments, there is not necessarily a loss of individual identity.[81] Communitas is dependent on 'being true to oneself,'[82] or for Turner, the 'authentic self,' beyond 'playacting.'[83] Individuals, then, have become a part of a collective of identities in which they find some sense of belonging and verification of themselves and this not only leads to a feeling of cohesion with group and place but of wellbeing through these stages of transformation.

Braaten suggests that within group cohesiveness comes a shared sense of wholeness, a gestalt, a whole that is more than simply the sum of its parts. The group is no longer a collective of individuals but seems to carry a sense of unity. With this kind of group cohesiveness comes 'collective intention,' this being a term coined by Tuomela and Miller,[84] alongside the phrase 'we-intention' or 'cases of joint social interaction' and often a joint goal. This concept is helpful when considering the activities. A basic collective intention may be that, for example, we will climb that hill, or it may be as open as we will have an *experience* (perhaps transcendental or mindful, fun, or based on hard work). This concept allows us to consider that as a group, individuals can have a collective will, like that of Gilbert's 'plural subject.'[85]

This could also be considered as Bratman[86] does 'shared cooperative activity.' For an activity to be described in this way, it needs to have appropriate behaviour, mutual responsiveness, goals, commitment to joint activity, mutual support, reciprocal expectations, cognitive interdependence, or at the very least an intention to join in an activity,[87] even if the agent has 'such an intention for different reasons.'[88] We know that each individual has differing agendas and motivations, and of course there can be clashes within case study groups of we-intentions and I-intentions. However, the group as a whole will begin to set up collective we-intentions once there is a sense of cohesiveness. Some of these intentions are imposed (e.g. by the organisation, to walk a certain distance or to sit in silence for a certain amount of time), and some are collectively decided (e.g. to have a positive experience or to be civil). Facundo[89] follows Braatman by going on to state that there must be a mutual reliance in shared intention and with this comes a relevant moral obligation: a collective obligation.

The group could be said then to have crossed another metaphorical liminal threshold. This time that threshold is within the dynamic shift from a collection of individuals to a cohesive group. This threshold seems to be crossed and uncrossed. It is not a definitively crossed threshold—as the group dynamics shifts and alters, individuals react and dip in and out of this cohesive state. With a collective form come upon through sharing in experience and developing the self and the group, there is a collective intention and obligation. I believe that this alters the groups' collective perception of the landscape in which they are in. This is to do with new-found contexts, new narratives, and new associations made within these spaces.

Proshanky et al.[90] suggest that

> place identity is developed by thinking and talking about places through a process of distancing which allows for reflection and appreciation of places.[91]

I will discuss concepts of place identity in relation to my case study groups when I consider the perception of the landscape, in relation to the self, as a liminal site.

The Perception of Physical Landscape as a Liminal Site

The facilitators, visitors to The Community Living initiative, The Loose Community, Woodland Weekenders, and The Mental Health Initiative participants met their excursions with already predetermined expectations that may, due to this, have been actualised in experience. The Young Adults, however, with little context came with little expectation other than that they may be challenged. This concept of challenge altered their expectations and thus experience. They also came with a knowledge of intentions of the project instilled by the facilitators. This also effected experience and perception of the activities and landscapes (Fig. 5.6).

For my case study groups, these experiences are both effected by an inexplicable innateness and culturally learnt behaviours, reactions, and perceptions. Some believe that these excursions bring about certain reactions due to an innate human relationship with nature, yet their ideas around our relationships with nature are culturally derived. For others, their relationships to nature come from past experiences, contexts, and knowledge or personal narratives. The way that a space is utilised and experienced depends on this 'knowledge' and narrative and the ability to abstract ideas. Ultimately this will have an effect on not only what is done (depending on agenda) but the experience of the spaces as a whole.

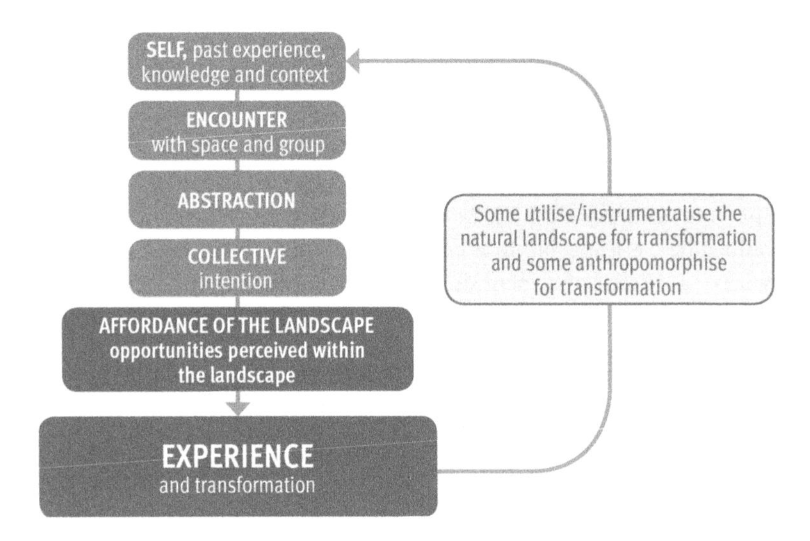

Fig. 5.6 Context, abstraction, affordance, and experience

For Carrier,[92] an abstracting agent can act purposive within situations; they can therefore shape the situation. This led me to Gibson and later Heft's 'affordance theory':

> The affordances of the environment are what it offers the animal, what it provides or furnishes, either for good or ill.[93]

> [The affordance of something is] a specific combination of the properties of its substance and its surfaces taken with reference to an animal[94]

According to Heft:

> The affordances of a given place in the environment establish for an individual what actions are possible there and what the consequences of those actions are.[95]

Whilst Gibson's earlier definition is rather open, his later definition allows little flexibility with regard to what an affordance could be. However, Heft's later definition, involving the possibilities of various actions and the potential for multiple consequences allowed me scope to explore a correlation between abstraction, affordance and/or *opportunity* and experience. To help with this theory, I first quote Heft and take his words as licence to perhaps push this further:

> much of the functional meaning in our perceptual experience is not of this nature; it is culturally-derived. And extending the concept of affordances to meanings that are specific to a culture may seem unwarranted, or at the very least, a careless and inconsistent application of the concept. Can the affordance concept be applied to cases of culturally-derived meaning, or is it to be limited only to those meanings of a more transcultural or species-specific nature? When affordances are viewed narrowly as body-scaled features, in the manner discussed above, then indeed the concept will have little to say about the culturally-derived meaning of objects. In this case, the concept would have interesting, but somewhat limited usefulness in tackling the problem of meaning in perception. *However, if affordances are specified relative to the individual's intentional repertoire, the prospects for a wider applicability are promising.*[96]

These landscapes, alongside the groups that enter them, provide an opportunity to do many things (physically) and to think about many things (mentally). Objects may afford specific action but context and understanding allow this perception of affordance. Perhaps then we may consider affordance instead as opportunity for social and material experience dependent on group dynamics, knowledge, and contexts. Context, knowledge, and abstraction lead to perceiving opportunity for types of experience and ideas of place within environments. This in turn may suggest that those are the types of experience that should be set as goals within those environments. If this might be the case, experiences then, are predetermined by the self (and personal context) and the group (i.e. predetermined intentions). If we think about abstraction then, an opportunity for action and experience is seen due to contexts and the abstracting of ideas, or it may be seen due to understood intentions of the group. The ability to seize these opportunities will have an effect on experience.

I am not as interested in the innate affordance of a rock or in 'body-scaling'[97] but instead the idea that a space may afford an alteration in way of thinking, socially interacting or perceiving oneself. I will say also that the perception of affordance/ opportunity is changeable throughout experiences—perhaps changed over the course of a day or perhaps weeks. These affordances, or opportunities, may always be present however not always immediately perceived: a canoe may afford an action; however, evidence shows that the experience of this action also affords a conversation, a new way of thinking about this action, and a changed perception of what it means to get into a canoe—not only to sit within a small boat and bob along the water but to put oneself in a challenging situation, conquer a fear, and perhaps leave feeling differently. The activity, location, and material object, alongside abstract ideas and intention afford the *opportunity* to think of oneself differently.

A person with the belief or the 'knowledge' that the non-human or natural can be a powerful force to their own self may anthropomorphise the non-human. They may appreciate the landscape for this end: to feel something. Someone without this context may not. Similarly, a person who 'knows' that climbing a mountain may have a positive effect may utilise the landscape for this end, using their own human intention to climb it. Again, someone without this context may not. Groups offer a physical and metaphorical space in which to build this context. They offer opportunity to expand this knowledge and insight into how a natural space may be afforded to have some kind of transformative effect on the self and on the group.

An affordance then, is what something may offer, not only physically but mentally, not only to the development of the self but also to the possibility of group interaction and group development: an opportunity. The openness to these opportunities is down to an individual's understanding or contextual framing of the space and circumstance. These contextual frames are first set up in the individual's life beforehand and secondly established by the organisation, group norms, group intention, and sometimes obligation. Abstraction enables going beyond the immediate perceived affordance—going beyond seeing that a hill may be climbed and, in some cases, seeing beyond the possibilities of merely human agency as a player within these situations. This is a going beyond the physical and physiological and therefore acting on more abstract notions of what the space allows and perhaps provides—this affects experience.

> Wild camping and things as well through going out and camping at market bay and just seeing, you know, the beauty of that place and spending

> over night there and being able to watch the sunset and enjoy the beach and, and things but also getting out and doing expeditions as well so people have extended time out, and you know and that develops things, like self-reliance and looking after each other and, and cooking for themselves and looking after themselves and making sure that the group works together as well. (Allan)

The Bay over the gulley is Allan's favourite place on the island. On my first afternoon here, Allan took me to visit along with his dog Tarfi. When we got there, Tarfi waited to pounce with seaweed hanging out of his mouth. I struggled to slide down the rocks that hung over the bay and on to the beach having already fallen on my behind in the mud. Allan had slipped twice and hadn't even acknowledged it. He walked on as if falling was the standard. He talked about the colours of the sand. Allan often talked about the colours of things: yellows, golds, and how in the autumn it is beautiful. Allan often swam in the sea and claimed that it was cold and cleansing in late December.

Later, when I returned in the spring I too swam in the bay and the water took my breath away before becoming warm. It was easy to see then why Allan might swim. Experiencing the sea physically completely altered the way that I looked at the bay for the rest of my time there. As we stared out across the sand and water, he talked about how the sunlight comes through the clouds on to the beach on an autumn day and how often he brought groups here to sit on the rocks and talk. For Allan, this space was a place of meaning—a sense built over four years of walking down to the bay and of encouraging others to appreciate what it has to offer. A new visitor, such as myself, of course may appreciate its aestheticism immediately, to *know* the place as Allan does, or to *know* Glen Etive as James does we need to bring the space into our own personal narrative (Fig. 5.7).

Building Context and Experience

> The [personal development charity] in the Gorbles, three years ago, I mean it was the first time they've been up here and there was about, I don't know how many, eight or nine of them and they spent most of the week going,

Fig. 5.7 The bay in winter

> 'nah, don't want to do that, nah, don't want to go kayaking, don't wanna go rock climbing, don't wanna go coasteering,' and sort of stood in groups around the edge of things, sort of watching things and when their leader took them back [to Glasgow], he was like 'Oh well d'you wanna go back [there]?' and they went 'yeah, yeah, we loved going' [...] [laughter] and he's going 'well you didn't do anything, you just stood around and said nut I'm not doing anything' and they went 'nah no, we loved it, we really enjoyed it, it was a really cool place to be, we definitely wanna go back.' (Allan)

The volunteers that Allan spoke about did come back to visit the island; however, when they returned, they behaved differently. The Young Adults wanted to take part, and they seemed to have more confidence in doing the activities. Allan believes that this is because they felt accepted here and that they got an 'enormous amount out of it' whether they fully participated or not. What this *amount* is, is yet again, intangible. For Allan, it is not about the taking part in activities; it is about 'experiencing each other and experiencing the place' and 'feeling safe.' What I think we are seeing within Allan's quote is that initial ambiguity, uncertainty, and fear often force people to watch apprehensively from the sidelines until they feel a sense of what is going on—what the frameworks and boundaries are. Though they may not have engaged in activities, they did engage with one another, and took in the space around them; they began to see themselves in the space and began to associate it as somewhere where they might belong. For Keith Basso having drawn on Heidegger and Sartre, there is a process of 'interanimation' occurring when people engage with natural places.[98] This happens when people:

> pay attention to a place, which causes it to generate its own field of meaning 'through a vigorous conflation of attentive subject and geographical object' [...] As the place animates the thoughts and feelings of the attentive visitor, these same thoughts and feelings animate the place in turn, in a reciprocal and dynamic process.[99]

My understanding of these situations is similar to Basso's in the sense that in being there and in engaging with others and the landscape the space is animated, it now suggests new opportunities to the groups and opportunities for positive experience. When people began to associate spaces with their own personal narrative, as no longer a place for *other* people but a space in which they may self-identify, their scope of engagement altered and in this, their perception of the space altered. For Tuan,[100] 'what begins as an undifferentiated space becomes a place as we get to know it better and endow it with value.'[101] This is in line with my supposition that through bodily attunement, action, physical engagement, reverie, and personal and group reflection, as well as through sharing these environments with other people and material actors, these landscapes become instead a series of *places*. Whether we are speaking of 'place-identity,'[102] a development of 'place attachment'[103] (to one specific place or place congruence for similar spaces),[104] or a developing 'sense of place,'[105] we are speaking of the relationship between the self and a physical space, of relating the self to the landscape.

For Bachelard, where Basso may say attentiveness, it is in fact the meditations that come from reverie that are 'indebted to the ways in which human beings invest meaning in, and give value to the world.'[106] Tuan[107] discusses that knowing a place both intimately and conceptually is one thing, but being able to articulate these notions as separate from the senses is difficult; most often because 'people tend to suppress that which they cannot express.'[108] Particularly if people do not have the ability to abstract ideas, as I have discussed. Kyle et al.[109] argue that individual identity is what makes a space a place of significance in that when a person can achieve self-verification from a space, attachment develops, I suggest that this self-verification in turn affects the dynamics of the group and thus yes, the perception of the landscape as a part of one's identity. Conradson[110] believes that the encounter between self and landscape is a relational one, in that there are felt dimensions within these encounters. The space has a significant effect on our beings. He too probes how being within a given landscape allows for a sense that subjectivities are altered, not only temporarily within the moment of encounter but indeed after the said event.

When subjectivities are altered and one's sense of self develops within groups, and when subsequently places inherit meaning for individuals, I agree with Kyle et al.[111] that reciprocally activities and social groups verify these place identities and 'are increasingly valued, further fostering individual's ties to place.'[112]

> as people develop attachments to significant others, the sentiment ascribed to those relationships also become intimately linked to the activities shared and the settings in which the relationship is nurtured.[113]

This comment furthers the notion of the cyclical nature of these relations between the self, group, and landscape as shown below in the figure '*The Liminal Loop*.'

> once you know where you are and you've been there a few times it almost feels like your just, your outside your back... almost in your back garden cos you know where everything is, you know the scenery, how harsh the landscape can be and you realise that oh, I know how to adapt to this situation [...] The heat plays a big factor so it can make you feel a bit grumpy, a bit sad, a bit tired but when you're up North in Scotland it's a bit miserable. It

> can be a bit more wet. You do get the sunny days where it changes off and on constantly. You've got to be able to adapt to that situation. Adapt to that landscape. (James)

James likens a familiarity with the natural outdoors, specifically in this case Glen Etive and the lands walked to reach this spot, to the familiar emotions that he feels when in his own garden. Clearly the two places are not the same—one being in the 'wild' north-west highlands of Scotland, vast with mountains, glens, farmland, rivers, and forest and the other a small 'back-green' in urban Edinburgh. For James, though, the outdoors is the outdoors regardless as both are, for him, natural. He claims that in revisiting spaces, you gain a knowledge of where everything is; he *knows* the scenery, and he *knows* the harshness of the landscape. Whilst of course knowing the scenery is an aesthetic experience, knowing the landscape is a visceral and psychological one. His knowledge of the space is built through gaining context within it—to associating it with his own personal narrative. This is experiential. He associates the weather with his mood and physicality. He anthropomorphises the north of Scotland as 'miserable.' He describes the weather as changeable and states that not only does he feel the need to adapt as a person to the situation but to adapt to the landscape. The landscape changes James. James' sense of self is extended to the space in which he stands and when reflecting on the landscape he equates it to himself visually, viscerally, and mentally. His relationship with the landscape is ever developing, and with each excursion to further natural spaces, he adapts and learns.

> Place and meanings of place stand forth as an ongoing process.[114]

The meaning that James associates with this place is an ongoing process of attendance, experience, and reflection in loop. He is encouraged to believe that he is the active agent over his experience of the outdoors and indeed what he may take away from these experiences into his everyday life. If the space changes in James' mind to become a place of significance and has an active effect on his person, then might we say that both James and the landscape are actors within his experience.

The glen near the bothy has become a *place* of meaning for James because he has taken part in activities there. He has built his own shelter in order to sleep outside there. He has played football at the foot of the hills. He has sat and reflected there. He has chatted to friends there. He has scrambled up the hills and returned red faced and excited before dinner there. He has walked there for hours and challenged himself to canoe across the river there. He has waded with wheelbarrow full of food and supplies in the rain there. He has chased new friends around the glen there and dragged wood from the loch to build with there. He has experienced there what he is capable of in a way that he has not elsewhere. In other words, James has related this specific place to his own personal narrative and in doing so has found a familiarity in this, no-longer-abstract space, but place of relevance to his own life.

The Dynamics of Experience

We see several things here. James has been witness to an emerging 'sense of place.'[115] At this site, he has become familiar with this place and self-identified with it. He has self-verified aspects of himself and has developed towards the ought self. (I say ought self rather than ideal here due to the circumstances of personal development and job seeking James places himself in.) This landscape has provided for James, and for others within the personal development charity, a threshold within the mind. This threshold sits between the perception of the landscape as unfamiliar, perhaps ambiguous, frightening, full of uncertainty, and the perception of the landscape as familiar and associated with oneself. The physicality of encounter, the aestheticism, and support of reflection and reverie allows the self to cross this threshold leaving one's perceptions altered. For James, the encounters with natural spaces with this case study were not his first. Whether for a participant they are gaining first contexts or nurturing ongoing relations with spaces, this landscape allows the mind to wander, the body to experience, the eyes to see, the group to share, and the self to transform.

The landscape can be considered as liminal in three ways: people can physically cross tangible thresholds—rivers, through caves. They can physically shape the landscape making claim to objects in ritual, crafting objects, and reconfiguring, managing, moving, and building within the landscape. The liminality can be seen in tangible and visible interaction. People also can be transformed—relational transformation of conscious and unconscious perceptions of the space (Fig. 5.8).

Natural landscapes have an impact on the individual and the group. This is relational; without the individual (and the group of which she has become part), the landscape could not be considered in the same way. If we refer to the figure above—as the self passes through liminal stages, whilst at the same time both influencing and being influenced by the group, alongside the group altering and passing through its own liminal stage (due to multiple self's merging and transforming) fostering a sense of communitas, the landscape begins to transform. This diagram shows this relationship as a continual cycle of transformation; with each new engagement with the space, the self, and the group, is affected. The 'self' changes, the 'self' changes the group, the group changes the perception of the landscape, and the landscape changes the self and thus the group again, and so on.

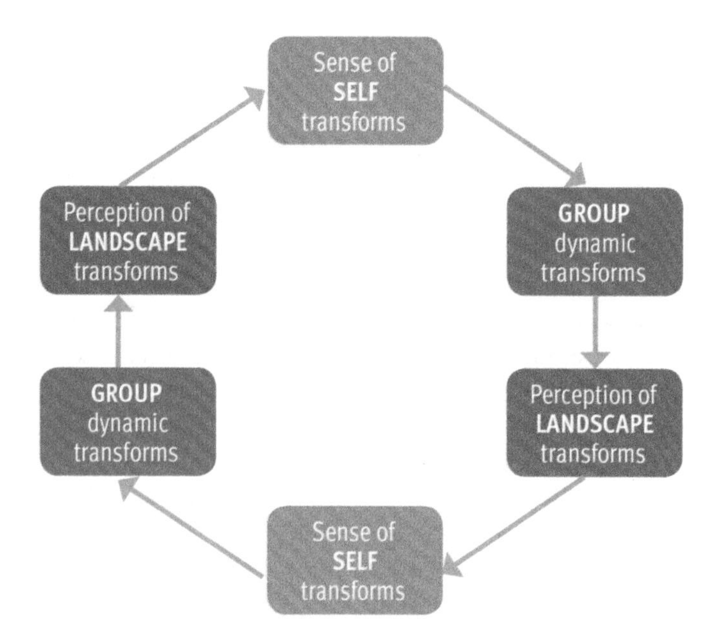

Fig. 5.8 The liminal loop—the dynamics of experience within natural landscape

Notes

1. Turner (1983).
2. Van Gennep (1960).
3. Turner (1967, 1969, 1974, 1982, 1983).
4. Gennep (1960).
5. in Turner (1974: 57).
6. Turner (1974, 1983).
7. Turner (1974, 1983).
8. Turner (1974, 1983).
9. Turner (1974: 86).
10. Turner (1974).
11. Turner (1974: 60).
12. Turner (1974).
13. Turner (1983: 64).
14. Turner (1983: 65).
15. Turner (1983).
16. Turner (1983).
17. Turner (1983).
18. Isiah Berlin in Turner (1983: 68).
19. Turner (1983).
20. as chastised by Thomassen (2012).
21. Turner (1967, 1983).
22. Turner (1974: 82).
23. Turner (1969, 1974).
24. Turner (1974: 82).
25. Olaveson (2001: 93).
26. Turner (1967).
27. Turner (1987: 105).
28. Lewis (2000).
29. Lewis (2000: 59).
30. In Lewis (2000: 65).
31. Lewis (2000).
32. Lewis (2000).
33. Turner (1983).
34. Beech (2011).
35. Braaten (1991).
36. Braaten (1991).
37. Braaten (1991).

38. Braaten (1991).
39. Noble and Walker (1997).
40. Tuckman (1965).
41. Tuomela and Miller (1987).
42. Tuomela and Miller (1987).
43. Facundo (2008).
44. Gilbert (1990).
45. Noble and Walker (1997).
46. Andrews (1999: 36).
47. in Andrews and Roberts (2012: 69).
48. In Andrews and Roberts (2012: 24).
49. Elliot (1999: 129–130).
50. Bakhtin (1965).
51. Turner (1969, 1970, 1974).
52. in De Azevado (2000: 83).
53. Bion (Ibid.).
54. Bion (1970: 87).
55. Bion (1970: 246).
56. Turner (1967: 106).
57. Turner (1967: 105).
58. Beech (2011).
59. Beech (2011: 289).
60. In Beech (2011: 290).
61. In Beech (2011).
62. Gilbert (1990).
63. Gilbert (1990: 07–10).
64. Bonebright (2010: 113).
65. Tuckman (1965).
66. Bonebright (2010: 114).
67. Bonebright (2010:114).
68. Tuckman (1965).
69. Bonebright (2010:114).
70. Bonebright (2010).
71. Bonebright (2010).
72. This may, on the surface, seem odd to an outside eye; however, in colloquial Edinburgh banter, to offer 'to batter' someone on another's behalf or indeed to say 'I will batter you' in many instances is not in fact as serious as it may at once appear. The term of course has the same meaning, but the sentiment behind it is entirely different.
73. Bonebright (2010).

74. Bonebright (2010: 14).
75. 'Performing' is the 'final stage of the original model' where the group develops 'functional role relatedness' (Tuckman 1965: 387). For Tuckman, the group becomes a 'problem-solving instrument' as members adapt and play roles that will enhance the task activities. Structure here would be supportive of task performance. Roles would become flexible and functional, and group energy would be 'channelled into the task.' (Bonebright 2010: 114) There is an additional stage that comes after 'performing' according to Tuckman. This was added later: 'adjourning' for Tuckman was added to the model in 1977, allegedly to reflect a 'group life cycle model in which separation is an important issue throughout the life of the group' (Ibid.) The leaving one another to return to the norm is, for Tuckman, the separation while for Turner coming together forming within the group would be the separation from the norm.
76. Braaten (1991).
77. Braaten (1991: 49).
78. Turner (1967, 1969, 1974).
79. Braaten (1991: 40).
80. In Andrews (1999).
81. Turner (1967, 1969, 1974).
82. Erickson in Andrews (1999).
83. Turner (1967, 1969, 1974).
84. Tuomela and Miller (1987).
85. Gilbert (1990).
86. Bratman (1992).
87. Bratman (1992).
88. Bratman (1992: 329).
89. Facundo (2008).
90. Proshanky et al. (1983).
91. in Brooks et al. (2006: 343).
92. Carrier (2001).
93. Gibson (1966: 127).
94. Gibson (1977: 67).
95. Heft (1989: 03).
96. Heft (1989: 12—emphasis my own).
97. Heft (1989: 03).
98. In Orley in Andrews and Roberts (2012: 37).
99. In Orley in Andrews and Roberts (2012: 37).

100. Tuan (1979).
101. Tuan (1979: 06).
102. Proshansky et al. (1983) and Twigger-Ross and Uzzell (1996).
103. Altman and Low (1992) and Milligan (1998).
104. Altman and Low (1992) and Milligan (1998).
105. Tuan (1979), Hay (1998a, b).
106. Picart (1997: 102).
107. Tuan (1979).
108. Tuan (1979: 06).
109. Kyle et al. (2014).
110. Conradson (2005).
111. Kyle et al. (2014).
112. Kyle et al. (2014: 1020).
113. Kyle and Chick in Kyle et al. (2014: 1024).
114. Gustavson (2001: 13).
115. Tuan (1979).

References

Altman, I., & Low, S. M. (1992). *Place Attachment: A Conceptual Inquiry*. London: Springer.

Andrews, K. (1999). The Wilderness Expedition as a Rite of Passage: Meaning and Process in Experiential Education. *The Journal of Experiential Education, 22*(1), 35–43.

Andrews, H., & Roberts, L. (Eds.). (2012). *Liminal Landscapes: Travel, Experience and Spaces In-between, Contemporary Geographies of Leisure, Tourism and Mobility*. Oxon/New York: Routledge.

Bakhtin, M. (1965). *Rabelais and His World* (trans: Iswolsky, H.). Bloomington: Indiana University Press.

Beech, N. (2011). Liminality and the Practices of Identity Reconstruction. *Human Relations, 64*(2), 285–302. https://doi.org/10.1177/0018726710371235.

Bion, W. (1970). *Attention and Interpretation*. London: Tavistock.

Bonebright, D. A. (2010). 40 Years of Storming: A Historical Review of Tuckman's Model of Small Group Development. *Human Resource Development International, 13*(1), 111–120. https://doi.org/10.1080/13678861003589099.

Braaten, L. J. (1991). Group Cohesion: A New Multi-dimensional Model. *Group, 15*(1), 39–55. Brunner/Mazel, Inc.

Bratman, M. E. (1992). Shared Cooperative Activity. *The Philosophical Review, 101*(2), 327–341. https://doi.org/10.2307/2185537.

Brooks, J. J., Wallace, G. N., & Williams, D. R. (2006). Place as Relationship Partner: An Alternative Metaphor for Understanding the Quality of Visitor Experience in a Backcountry Setting. *Leisure Sciences, 28*(4), 331–349. https://doi.org/10.1080/01490400600745852.

Carrier, J. (2001). Social Aspects of Abstraction. *Social Anthropology, 9*(3), 243–256.

Conradson, D. (2005). Landscape Care and the Relational Self: Therapeutic Encounters in Rural England. *Health and Place, 11*, 337–348. https://doi.org/10.1016/j.healthplace.2005.02.004.

De Azevado, A. M. A. (2000). Substantative Unconscious and Adjective Unconscious: The Contribution of Wilfred Bion. *Journal of Analytical Psychology, 45*, 75–91. Oxford: Blackwell Publishers Limited.

Elliot, S. (1999). Carnival and Dialogue in Bakhtin's Poetics of Folklore. *Folklore Forum, 30*(1/2): 129–139. Available at http://obook.org/amr/library/carnival_bakhtin.pdf. Accessed 27 July 2017.

Facundo, A. F. (2008). *Shared Intention, Reliance and Interpersonal Obligations: An Inquiry into the Metaphysics and Interpersonal Normativity of Shared Agency.* Stanford University, ProQuest Dissertations and Theses.

Gibson, J. J. (1966/1983). *The Senses Considered as a Perceptual System.* Santa Barbara: Praeger.

Gibson, J. J. (1977/2017). The Theory of Affordances. In R. Shaw & J. Bransford (Eds.), *Perceiving, Acting and Knowing an Ecological Psychology* (Kindle Edition, pp. 127–143) New York: Routledge.

Gilbert, M. (1990). Walking Together: A Paradigmatic Social Phenomenon. *Midwest Studies in Philosophy, XV*, 1–14.

Gustavson, P. (2001). Meanings of Place: Everyday Experience and Theoretical Conceptualisations. *Journal of Environmental Psychology, 21*, 5–16. https://doi.org/10.1006/jevp.2000.0185.

Hay, R. (1998a). A Rooted Sense of Place in Cross-Cultural Perspective. *The Canadian Geographer, 42*(3), 245–266.

Hay, R. (1998b). Sense of Place in Developmental Context. *Journal of Environmental Psychology, 18*(1), 5–29. https://doi.org/10.1006/jevp.1997.0060.

Heft, H. (1989). Affordances and the Body: An Intentional Analysis of Gibson's Ecological Approach to Visual Perception. *Journal for the Theory of Social Behaviour, 19*(1), 1–30.

Kyle, G. T., Jun, J., & Absher, J. D. (2014). Repositioning Identity in Conceptualisations of Human-Place Bonding. *Environment and Behaviour, 46*(8), 1018–1043. https://doi.org/10.1177/0013916513488783.

Lewis, N. (2000). The Climbing Body, Nature and the Experience of Modernity. *Body and Society, 6*(3–4), 58–80.
Milligan, M. J. (1998). Interactional Past and Potential: The Social Construction of Place Attachment. *Symbolic Interactionism, 21*, 1–33.
Noble, C. H., & Walker, B. A. (1997). Exploring the Relationships Among Liminal Transitions, Symbolic Consumption, and the Extended Self. *Psychology and Marketing, 14*(1), 29–47.
Olaveson, T. (2001). Collective Effervescence and Communitas; Processual Models of Ritual in Emile Durkheim and Victor Turner. *Dialectical Anthropology, 26*, 89–124.
Picart, C. J. S. (1997). Metaphysics in Gaston Bachelard's "Reverie". *Human Studies, 20*(1), 59–73. http://www.jstor.org/stable/20011137
Proshansky, H. M., Fabian, A. K., & Kaminoff, R. (1983). Place Identity: Physical World Socialisation of the Self. *Journal of Environmental Psychology, 3*, 57–83.
Thomassen, B. (2012). Revisiting Liminality: The Danger of Empty Spaces. In H. Andrews & L. Roberts (Eds.), *Liminal Landscapes: Travel, Experience and Spaces In-between, Contemporary Geographies of Leisure, Tourism and Mobility*. Oxon/New York: Routledge.
Tuan, Y. F. (1979). *Space and Place*. London: Edward Arnold Publishers Ltd.
Tuckman, B. W. (1965). Developmental Sequence in Small Groups. *Psychological Bulletin, 63*(6), 384–399.
Tuomela, R., & Miller, K. (1987). We-intentions. *Philosophical Studies, 53*, 367–389, 1988.
Turner, V. (1967). Betwixt and Between: The Liminal Period in Rites de Passage. In *The Forest of Symbols*. New York: Cornell University Press.
Turner, V. (1969). *The Ritual Process – Structure and Anti-structure*. Suffolk: Richard Clay (The Chaucer Press Ltd).
Turner, V. (1970). *The Forest of Symbols: Aspects of Ndembu Ritual*. Ithaca: Cornell University Press.
Turner, V. (1974). *Dramas, Fields, and Metaphors*. Ithaca: Cornell University Press.
Turner, V. (1982). *From Ritual to Theatre*. New York: PAJ Publications.
Turner, V. (1983). Liminal to Liminoid, in Play, Flow and Ritual: An Essay in Comparative Symbology. *Rice University Studies*, 53–92. Available at https://scholarship.rice.edu/bitstream/handle/1911/63159/article_RIP603_part4.pdf viewed on 27.07.17. Accessed 30 Aug 2017.
Turner, V. (1987). *The Anthropology of Performance*. London: PAJ Publications.
Twigger-Ross, C. L., & Uzzell, D. L. (1996). Place and Identity Processes. *Journal of Environmental Psychology, 16*, 205–220.
Van Gennep, A. (1960). *The Rites of Passage*. Chicago: Chicago University Press.

6

Anthropocentrism, Agency, and the Transforming Self

The previous chapters have outlined what natural space represents for informants in terms of why they may choose to engage with these social and physical spaces. As we have seen in Chap. 4, the motivation to engage with groups and in natural space appears to be due to a desire to belong within a given group and to verify some sense of the self in doing so. This is an aim to verify an aspect of one's identity, found by informants within the group's ethos and in the group's activities. We saw that the sense of self is enabled to metaphorically move closer to people's own perception of their ideal or ought self. With this, as we saw in Chap. 5, the sense of self becomes a liminal threshold, and when each person is transformed along these lines, the group dynamics changes, and ways of perceiving (and thus behaving within and towards) the landscape is altered.

Within my case study activity and in their agenda, I have alluded to seemingly contrasting and diverse ideas of how people engage with the natural landscape and for what means. To simplify, I say that these aims are evident from one end of a, albeit complex and perhaps jumbled, spectrum to another. On one side of this spectrum, the aim is to go out in nature and challenge oneself physically and mentally utilising one's *own agency*. On the other, the aim is to go out in nature, to communicate

R. Crowther, *Wellbeing and Self-Transformation in Natural Landscapes*,
https://doi.org/10.1007/978-3-319-97673-0_6

with the non-human in a *reciprocal* manner—thus, allowing nature to influence you as mediator of the experience, that is as an *active agent* within experience.

Individual relationships with the non-human within these natural spaces were complex, and thus these relationships seemed to have a complex effect on experience. Here, I now explore how and why informants seemed to align themselves with one way of thought rather than another—why some informants focus on their own efficacy and others on the power of the non-human in experience.

First, I will deal with the later end of the spectrum—the perception of non-human agency as mediator in experience. Later, I also question whether informants are indeed aware of their own positioning. In doing so, I question whether there was a sense of jockeying for moral position within the larger culture of outdoor engagement and consideration of the natural landscape. With this, I refer to individual's beliefs regarding what it means to be a *good person*, or to have a sense of oneself as *doing good* within, alongside, and towards the landscape, as intermediary or mediator of experience. Whether this may be in empathising with nature or in preserving or conserving nature, these alignments could be said to influence wellbeing. I will discuss what it means to group members to do 'good' in relation to nature in Chap. 7. This brings us back to issues relating to self-identity and the desire to belong to groups that may self-verify one's ought or ideal sense of self. It also brings us back to ideas of abstraction and the effect personal narratives and agendas have upon experience of the landscape. At this point, I will outline the inevitability of centrism.

The first factor in where a group's perception lies is the overall agenda and philosophical (or indeed pragmatic) ethos of the group. The second factor is the development of individual perceptions, personal narratives, and abstract ideas of an individual's relationship with nature. These emerge via the developing sense of belonging and the group's experience as a dynamic whole. The group's influence on one another is seen to influence the anthropocentric attitude within these excursions, whether in considering the human as the only active agent, to personification, and anthropomorphic perception of the non-human. We can approach this through trying to comprehend behaviours, human values, language, and metaphors used by individuals. We will also see these shifting dynamics within the group and the way that the space was considered as liminal. When this was observed and subsequently analysed, it became evident that the ways in which people think about, approach, and behave within and towards the landscape is multifaceted. Perceptions of the natural landscape among the groups entail seemingly opposite philosophies—either anthropocentric or ecocentric. However, when I looked further this was not as straightforward as it seemed.

> Other philosophical approaches might seem to confer a greater cogency, or even inescapability, on anthropocentrism. Thus, one version of anthropocentrism is simply the thesis that we, as human beings, cannot help making all our valuations with human faculties and from a human perspective. Frederick Ferré has named this harmless claim 'perspectival anthropocentrism.'[1]

Whether one projects human values upon the non-human metaphorically, or believes in only the agency of the human over the non-human (believing that our relationship is purely aesthetic or instrumental), or whether one considers the non-human as having not only intrinsic value but social communicability, the group's different relationships to the landscape are complex and contradictory. Might all participants be deemed as 'perspectival anthropocentris[ts]'[2]? When analysed, these seemingly dichotomous ways of thinking began to unravel and entwine. When coupled with theoretical enquiry and variations on terms and meaning within and around what it is to be anthropocentric, I began to unearth what it was that people wished to achieve. I began to understand what they wished to feel *a part of* and what effects these different perspectives had on overall experience. I considered the way in which the groups thought about their experience in terms of who or what was at the helm of initiating experience. We can understand the motivation to take part in these excursions when we consider what actors are perceived to have agency over these experiences: what is intermediary and what is mediator? By exploring this, we can understand the effect of these perceptions within time spent in natural landscapes.

Reciprocity, Personification, and Anthropomorphism

On my way to meet The Loose Community very early one morning, I sat at the back of a double-decker bus, anxious that I had no clue when to get off. I was to meet Finlay in the car park of a branch of Lidl's in Dalkeith. I counted each stop whilst I watched the tracker on Google maps on my iPhone. He arrived in a beaten-up Peugeot from the early 1990s. We were heading to Wiston Lodge in Biggar, South Lanarkshire, for a day of mindfulness and Goethean observation.

We gathered on the gravel outside the lodge and waited for everyone to arrive; everyone in the group arrived having car-shared with someone else. We headed into our stately living room base for the day. Some people sat on the Persian-rugged floor and some on the rather dated sofas in this former Victorian hunting lodge. I sat closest to Helen, and we waited for Ryan to arrive. Ryan is a Goethean scientist and community practitioner. He described Goethe as 'Germany's Shakespeare': a poet who was interested in the scientific and who related it to the natural world:

> [Goethe] was interested in what you can see and touch but also on the emotional aspect of seeing. It's about bringing in both aspects of experience. And to relate the non-human in a new way. To consider the tangible and the invisible experience. It is a mindful practice, a process that pays heed to focus and diligence. (Ryan)

Ryan's intention is to bridge the gap between 'what we can't see and what we can.' He believes it is an 'empathic practice and can be extended to the more-than-human.' Through imagination we can 'know' the non-human, he claimed. Ryan aims to 'bridge the gaps between social perception and natural processes.'

We began the morning with a 'mindful circle,' where we were encouraged to extend our peripheral vision and to look with a 'soft gaze.' This is meant to allow for an unfocused awareness of our entire surroundings and an observing of our bodies in relation to the floor, the air, and the sounds all around. We stood on the forest floor beneath a variety of trees, and the ground was crunchy with beech cones and twigs. It was a warm day, and there was an overwhelming crescendo of crows cawing above. We were given direction—to consider the elements and be aware of our surroundings but to simply walk around the forested paths of the grounds, without adhering too closely to these instructions. I wondered at this moment about whether the term mindfulness was perhaps synonymous with tenderness—tenderness for the moment, yourself, and the space in which you are in. The aim was to discover 'more than what [was] sense perceptible.'

> Dynamism of perception emerging from participating with the world around me, breeds dynamism for life itself. It is invigorating and enlivening and although at times challenging [...] the process is one of emerging mutual reciprocity and enhancement. (Ryan)

Ryan was asking us to 'just be' within the woodlands. To wander with a soft gaze and be aware of ourselves in situ. To be *mindful*, one must acknowledge when the mind wanders beyond the now, perhaps to deadlines, emails, and dinner, and to bring it back to the immediate senses in the moment. Ryan, however, also asked us to explore more than what was sense perceptible. What did he mean by this? My hunch was that the senses of which he spoke may not be our own.

The notion of wellbeing for this group seems to come from the sense of being enlivened and aware of material surroundings: 'In-touch.' Perhaps this is a moral standpoint in which we must not only focus on within ourselves but on the outside world, what it can offer us and what we might, in being enlivened, offer back to the natural world through our awareness. It is an enchantment of the everyday, that is, reciprocal between the human and non-human entities within a space. For Ryan, perception itself is dynamic and comes about through an active participation in the world which in turn renders life itself as dynamic: an enlivening, challenging, and mutual process. He believes that both the human and the non-human can benefit from this relationship. We might surmise then that by sensing the life within the non-human enlivens the group. But it seems that this 'life' is one of human value.

After our solo walk, we spent one hour observing a sycamore tree. We stood about 20 yards from the base of the tree and gazed up. It was an *elderly* tree, or should I say, a tree that had been growing in that spot for at least 100 years. The instinct to personify is not lost on me. Is it innately human to do so? We were asked to shout out comments regarding the tree, but we were specifically told that these comments must be descriptive and factual and must not be impressionistic. This seemed to be a struggle for these individuals. The notion of *presence* seems to be dominant. Not a single person shouted out the obvious—it was a sycamore tree by name. I wondered if I had missed something as I refrained from doing so. Instead there were distinctly anthropomorphic terms:

> Kindly! …Friendly! …Shy!… Humble! …Quietly confident! …Austere! […]

When we moved closer and sat beneath the sycamore, we were asked to look slowly upwards, observing section by section of the tree, commenting again on what we *saw.* The group again seemed to have difficulty in not being impressionistic nor metaphorical. Instead, they played in simile and abstraction—it looked like a 'human neck,' it's 'skin [was] cracked,' and it was like 'an armpit.' The group related the tree, aesthetically, to the human and then too, to the animal world—it was 'furry,' 'feathery,' 'fluffy,' like 'a crocodile.' We spent some more time closely observing a younger version of the tree. This was an attempt by Ryan to allow us to comprehend the growth and development of the tree from seedling to sapling to the elder that we had seen before us. The group noticed the sapling sycamore's 'baby leaves' and that its newly forming branches made it seem like an 'awkward teen.'

This group intends to see the value of the non-human as intrinsic. However, even with this agenda, could not avoid the trap of metaphor and enforcing human value. One could argue that as a group predominantly made up of creative types—writers and artists—the difference between descriptive and interpretive may have been lost, but I believe something else was at play here.

After three hours of observing the trees, we were asked to sit or lie down on the lawn outside of the lodge, where the giant sycamore rose above, and to close our eyes. Ryan had asked us to imagine the growth of the tree, from seed to full grown. We were to imagine it 'growing, stretching' and 'the energy and strength' it would require doing so. The last thing that we did was work with art materials to create something 'representative' of the day.

They said I was like
Like so many human centred things of their imperfect understanding
Sometimes I just was
Was in their human minds, their best efforts to understand
Then when they tried to be me
Me tree
I stretched their minds
Beyond earnest speciesism
Into my life force, pushing upwards, outwards
Stretching towards the sun
The same sun they turn their faces up to
The same life force (Jill)

Christie drew the Holy Trinity, represented in three deep black charcoal lines and described them as the roots beneath the tree. She described these roots as *like* the social and cultural *roots* that we have as human beings in society. These, she said, were the roots that define our human relationships and give us strength, as too they give the tree its strength. Another participant asked us to consider how many times it had been cut down or 'injured' and had been required to regrow. She mentioned rebirth, again drawing similarities to our necessity to grow as people through suffering and being metaphorically 'cut down.' Jill wrote the poem that you see above. This was a more explicit anthropomorphising of the tree. Perhaps, though, rather than strictly anthropomorphising with conviction, she was using metaphor to expose the contradictions in how we could possibly understand the non-human. She gave the tree a human voice whilst at the same time problematising our associating the tree with human attributes. She was attempting to 'articulate' the inarticulable—an attempt to sense the tree in the way that Ryan had wanted us to. For Rautio,[3] articulation as utilised in critical geography and anthropology[4] 'refers to an act of giving verbal form to nonverbal, elusive or subtle phenomena'; however, she also considers that

> to keep trying to fit world into words is to work with one's relations to the surrounding world. It is not to reduce world into words, or to agonize over a correspondence between the two.[5]

Prior to this excursion, Jill and I had discussed the cultural influence of language on how we perceive and relate to the 'natural' world. Jill questioned the ease of which we may possibly 'sit around in nature' without 'having our organising thoughts dictat[e] what [it is] you are trying to do.' She decided that it is an impossibility to allow nature to 'speak' to you whilst at the same time working along the premise of going through meaning-making activities designed by you. For Jill, this incapability adulterates and mediates her experience. In order to be completely blended sensually with the environment, in order to allow the non-human the power of mediation, we must first 'be nothing,' for some achievable through meditation, in an environment. One must have had no human conditioning, nor culture, nor understanding of language.[6] Without these conditions, humans cannot but attach symbolism, at the very least, to their environment. This returns us to Ferré's concept of 'perspectival anthropocentrism.'[7]

Jill told me about an excursion to Knoydart with Helen and some others within The Loose Community and a time that she was aware of herself being unable to 'allow nature to be itself.' On her first solo walk on this trip, a common exercise among this group, she had come across a flock of sheep. She thought to herself that they were 'a real chrome yellow colour.' This was because they had been dyed by the local farmer. Farmers do this, apparently, to have a system of knowing which sheep have been mated and which have not. 'These sheep were totally orange! They really stood out, they looked surreal, they looked like hyper sheep!' Jill laughed. She then told me that she had picked up on this colour everywhere on the island—in the lichen stuck fast to the dark grey and wet rocks, in plastic found on the beach, and in the flowers. She could see the contradiction and difficulty with the concept of perceiving nature with intrinsic value when the only language that we have to describe this phenomenon is 'most definitely human':

> and of course, what do I do with this [the colour]? I decide that this was a motif for cowardess [sic.] right? [laughter] and I write something about

that [...] *So* I'm in this thing where I'm constantly making meaning for things, seeing patterns, letting nature speak to me but I'm not really, I'm speaking to it and you know [laughter] so it's that dialogue that is quite challenging [...] I spent ages trying to describe the proper colour and I decided to call it chrome yellow because yellow wouldn't have been right, it had to be chrome yellow [making fun of herself] [laughter]

Jill is aware that if we were to truly approach the non-human world, without applying human value, the colour wouldn't in fact have been a *colour* at all. We certainly wouldn't be further associating cultural symbolism such as chrome, with the connotations this carries, of urbanism and human influence, with the landscape. Ryan believes that in the naming of things within the non-human natural world he begins to unnecessarily categorise which prevents him from being able to relate to things in the way that he would like to. For Ryan though, it is important to be able to communicate with each other about what we are seeing and experiencing, and without language we would be unable to do this:

I think in terms of that relationship with trees though I think very quickly we can just label that sycamore therefore we don't need to look at it we don't need to, you know that's not really reading the language if were gonna call it that, that's just labelling something and yeah in social environments we all know that labelling things can get us into trouble [it] limits the kind of conversations we can have and limits the kinds of interactions we can have with people or things and I think that's part of the challenge I suppose, is to get beyond that... sorry just to go back about the song that you were saying,[8] yeah I think we do know things from a different place and I think labelling things stops us from actually interacting with that notion, I don't know the deeper knowledge, a deeper intuitive knowledge [...] and I think that's what Goethean science is about, getting to that place [...] revealing things that were maybe always there but you didn't know were there so it's not necessarily that agency, although agency helps us get there I think it's there before we even start doing that to our agency. It's there because, because relationships are there. There's a lot more that we humans, at least in this culture, we don't tend to access, we don't tend to develop or aren't generally aware of it [laughter] and there's a lot of richness in there that I guess, you know, we haven't unravelled. (Ryan)

For Ryan, unlike Jill, there is a perceivable 'language of nature.' Ryan believes that throughout his Goethean observation and mindfulness workshop, the group

> got a sense of what the sycamore tree [was] and [...] a little bit of what's specific to the Sycamore tree as opposed to the oak tree or the birch tree.

He believes that each species in nature has its own characteristics and its own qualities and that these elements are an expression of 'something' (there is that *something* again!). This *something* can, for Ryan, be a form of communication and therefore a language. If we were to spend the same amount of focused time with an oak tree, for Ryan, we would see and understand different things—'through our sense[s] and through the way we engage with it.' He believes either an oak or a sycamore will have differing effects on the group and that this is a *language*. From my understanding, this could actually instead be put down to further cultural connotations of what an oak tree or otherwise *is*—the personification of a characterful oak tree as the king of the forest or the silver birch as the queen, for example. Whilst Jill uses metaphor and linguistic creativity to attempt to understand her relationship to nature as partly cultural, Ryan seems to disregard the influence of culture upon his relationship.

For Epley et al.,

> strong forms of anthropomorphism entail behaving as if a nonhuman agent has humanlike traits or characteristics along with explicit endorsement of those beliefs [...] whereas weaker forms may only entail "as if" metaphorical reasoning.[9]

Thus far, these people seem to have been displaying this weaker form of anthropomorphism, playing with language. They adopt simile and metaphor and place human value on to the sycamore tree. We have not yet seen anyone who explicitly experiences the tree as having communicability or intention. This is a mode of thought that appears to come only from a few of whom I spoke to, specifically Helen and Finlay who both are practising neo-shamans. Helen's and Finlay's contribution at the end of this workshop suggested that not only do they seek the same language that Ryan searches for but that they also perceive the non-human as agentic. Finlay chose to read us a section of his written reflection, a letter to the sycamore tree:

F: [I've] only just scratched the surface of knowing you, and what it's like to be you [...] I am left with a sense of longing [...] [it's] not just an experience and then home, today has brought up so many things in my life.
R: Longing is important as it's part of belonging. (Finlay and Ryan)

Finlay suggests that there is a way to *know* what it is like to be the tree and that this is part of a process that he had been unable to achieve on this day's workshop. Not only does Finlay believe in the possibilities of *knowing* the tree, he also associates this dialogue with his own personal narrative. This is further emphasised by Ryan's comments associating the longing that Finlay feels with *belonging*. For Finlay, this sense of a desire for belonging that is found within all my case study groups is, rather than wanting to solely belong within the group, referring to achieving a sense of belonging in relation to the non-human. This cannot fail to conjure the old adage of 'feeling at one with nature.' I feel within this research however that such complexity may not be understood with such a bromide.

> The projection of human attributes onto non-human domains is often explained in anthropology as the consequence of a tendency to animism and anthropomorphism present from the earliest stages of cognitive development.[10]

Helen chose to draw 'the energy of the tree growing' and had imagined different stages of its growth as a series of sounds, of course only explicable as human vocal sounds. She talked us through her drawing—squeaky pops for the buds bursting through the flesh of the plant and moaning groans as she imagined the trunk stretching wider for the group. The climax of development was a gentle humming harmony representing the canopy blossoming and spreading. When asked what the harmony may sound like, she hesitated before closing her eyes and improvising a floating and high-pitched melody.

I feel that Helen may be unique within my research in her perception of the non-human world, however perhaps representative of a few. When I spoke to Helen at my home before this excursion, she told me about a text, Jane Bennet's *Vibrant Matter*[11] and of how she resonates with the ideas within. For Bennet 'stuff' exhibits 'thing power'—it *speaks* to her, it produces effects—objects are 'vivid entities not entirely reducible to the contexts in which (human) subjects set them.'[12] Additionally,

> Bennett's (2001) insistence on the power of nature to enchant and nurture a spirit of generosity which is a necessary bringer-to-life of wider ethical codes. She contends that 'enchantment is a mood with ethical potential' which can 'aid in the project of cultivating a stance of presumptive generosity (i.e. of rendering oneself more open to the surprise of other selves and bodies and more willing and able to enter into productive assemblages with them)'[13]

Earlier I said that Ryan's insistence was on being aware and receptive to the non-human and that this was perhaps a moral issue. Here again, through Bennet's ideas, and Jones and Cloke's deconstruction of them (ascertaining that the ability to enchant the non-human has ethical potential, therefore, a potential that may allow for 'enter[ing] into productive assemblages with them') suggests that Helen's line of thought is similar to Ryan's, though more strongly anthropomorphic. For this reason, I feel, though Helen's beliefs are a slight exception within the group, they are certainly not without merit nor are they individual on the wider scale. Helen considers all material non-human, not only living natural things, as having a vibrancy constituting to agency and often intention:

> I was sitting in the kitchen through there and I was just looking at the light shining through that turquoise-y coloured glass and I was looking at the rainbow thing on the window and there was a really beautiful mug, sort of curvy mug with sort of patterns […] and all sorts of things, they just felt so, they seem to have so much power of their own and I wonder if […] I feel a lot less inclined to differentiate now between what is known as natural world and the kitchen for example. There's a kind of vibrancy, aliveness in all kinds of places and I think *that* probably develop*s*, it depends a lot more on how open I am to it… (Helen)

Helen explicitly states that she has more empathy with living organisms and that she considers living things as having more soul but that 'all sorts of *things* feel'; however, she can't 'know' if this is the case. As a child Helen felt that physical objects felt pain, not only the representations of living things, 'like dolls and stuffed animals' but also tables, chairs, and stones. She wondered whether this was residual in her or whether she had been socialised out of these thought patterns. Knowing Helen, I don't feel she has been. Perhaps she referred to others as having been socialised out of it rather than herself. Helen is a curious person, always championing the worth of play and exploration, perhaps nostalgic for a childhood spent without conditioning—able to sense the essence and vibrancy of the non-human (Fig. 6.1).

Fig. 6.1 Helen holding a made object, ready to toss it in the river as an offering

The Inevitability of Centrism

For Cocks and Simpson,[14] there is a distinction to be made between 'weak anthropocentrism' and 'strong anthropocentrism,' and somewhere in between we have ecocentrism and non-anthropocentrism.

> Strong anthropocentrism is a perspective wherein the nonhuman environment is typically taken to be a commodity [...] Here, nature does not appear to have intrinsic value, and its extrinsic value is limited to its obvious contributions to humankind.[15]

Similarly, Attfield deems anthropocentrism to be

> the belief that nothing but human beings has moral standing or should be taken into account in ethical deliberations; this belief is also sometimes known as 'normative anthropocentrism.'[16]

'Weak Anthropocentrism' on the other hand is

> a perspective whereby the natural environment is still measured by its value to human beings, but the value is significantly expanded beyond the commodities associated with strong anthropocentrism. Nature allows humans to grow in character and to feel good about themselves by interacting and caring for something outside themselves.[17]

Further terms are ecocentrism and non-anthropocentrism, the former coming from John Baird Callicot, an American environmental philosopher. Ecocentrism is 'a perspective whereby nature has intrinsic value independent of its direct value to human beings.'[18] Some of the people I journeyed alongside may place themselves within this category of philosophical belief. For Hargrove, 'non-anthropocentric' is when 'human valuers value things other than themselves, and it is intrinsic, because human valuers value these other things for their own sakes.'[19] Perhaps non-anthropocentrism could be describes as 'ethical mindfulness,'[20] where one might be 'extend[ing] the idea of intrinsic value and rights' to the non-human and having a 'recognition of human embeddedness in co-constructive relations with the non-human world.'[21] Trees are given a degree of this recognition by not only The Loose Community but a care also provided by The Woodland Management and Mental Health Initiative.

The concept of the non-human as being a potential mediator within experiences is not limited to the shamanic practitioners within The Loose Community. Whilst, yes, it seemed to be more common for individuals to personify or consider the non-human using metaphor, there are instances of people from different groups holding similar implicit beliefs to those within The Loose Community.

Centrism, in this instance, is about perspectives and the subsequent values that are then placed upon the non-human within these scenarios (and indeed in life ethos). To be anthropocentric is to fail to see things from non-human perspectives. By extension, it refers to the results of this and in this case the way that the landscape is experienced and spoken about. Some value the non-human and natural environments only for their instrumental use, some imagine the intentions of the non-human as like that of the human, and some use metaphor pertaining to human understandings of experience to talk about the non-human. There are also individuals who consider the non-human as communicative, of course in the only way that we might understand communication—as bodily, sensual and linguistic. Again, this is still anthropocentric.

An implied notion of reciprocity between human and non-human actors is seen in some of the comments made by Arnaud, a member of The Mental Health Initiative. His comments are stilted. He struggles to articulate exactly what his perspective is:

> I was wondering I mean we are part of the natural world, and I think that there is just a natural connection with nature which, I don't know if it's something that... I think its mutual. There's just something that...it's not coming just from me thinking I'm just going to go in nature because that's what... I think it's a mutual thing. (Arnaud)

Arnaud could be said to be being weakly anthropocentric or non-anthropocentric in that although he intends to value the non-human for its intrinsic value and displays an element of respect or awe, the reciprocity of which he speaks seems to be rather one-sided. Arnaud senses 'something'—that phenomenological sense of mutuality between nature and himself. He senses 'something' 'coming' from nature. Whether this is in an overtly communicative manner or not is not explicated. It is not, for Arnaud, a simple case of being in nature, but rather that 'something' more than simply being in the presence of the non-human is going on. For Arnaud, whatever this is, it ultimately leads to positive feelings in nature—his own subjective feelings of well-being. I wonder then if this is a social relationship.

If we return to Ryan and his notions of a reciprocal social relationships with nature, we may be able to consider what is occurring for these people.

Nature as Social

Ryan admires nature: he admires 'nature's ability to not reflect on life as social.' Does Ryan mean to imply that nature is capable of reflection? For humans, life and living is about perception. For Ryan, however, nature does not reflect (a thought process) on this perception, implying that it may reflect on other issues. Ryan believes that nature simply 'does living really well.' His comment was accompanied by laughter. What an odd concept he had stumbled upon! In Ryan's view, humans allow thought processes to 'get in the way,' and it is necessary for us to find a new way of relating to one another and our environment, 'not by reflecting and not by thinking about things but by focusing our reflections and thoughts in particular ways.'

Rebecca: you said that nature does living life very well, would you say that in some way nature is social?
Ryan: eh, yeah I think, social is nature maybe more
Rebecca: social is nature, OK, can you explain that?
Ryan: […] I think there are natural processes in social processes and all social processes have their natural process as well. I think nature's all about process, life is just actually all about process because everything is constantly in flux and changing

> and going through transformation and change. I think there are social dynamics to that within nature [...] I'm interested in how processes occur there and how we can, we can reflect on those in terms of how we live our lives or how we lead our social processes whether that's within organisations or our daily lives and try and live a bit more like nature [...] If we can bridge that gap a bit more then I think we would live in better balance and better relationship with nature in the bigger sense. (Ryan)

Ryan sees natural processes within social processes. He also believes that all social processes have natural processes. If I break this down, it seems that he means to say that within nature, whether in human nature or non-human nature, there is a sense of the social. And, within the social, there are innate processes gone through by both the human and the non-human. For Ryan, then, the natural and the social may be interchangeable. He speaks of transformation and a sense of social dynamics within the natural. He wants to understand the social processes that are incurred within the natural and in doing so reflect upon them alongside the human social process. Ryan wants to do this to be able to adopt nature's social processes into the way in which he lives his life. By doing this he intends to not only better his own social relationships with other humans but to bring about a balanced and better way of relating with nature.

Ryan's intentions, working alongside the perceived intentions of the non-human may seem drastic to some. However, it is not unheard of, particularly within eco-philosophy, where it is often assumed that people can or will attribute to nature the properties of the human mind.

For Ellen:

> There is nothing particularly mysterious about this [...] It happens because we are bound to model our world directly on those experiences which are most immediate, and these are experiences of our own body.[22]

What else, if not our own senses, body, and mind, would enable us to form any other relationship with the non-human natural within these landscapes? For those who wish to understand their relationship to nature is this anthropomorphism inescapably human? For some, as we have seen, this is more loosely based on personification, without the conviction perhaps of Ryan, Finlay, or Helen.

For Charles,[23] 'the quintessential mental trait is intentionality'; so if the non-human is perceived as intentional then they must also be perceived as having mental capacity. This is articulated often by The Loose Community. However, Charles also suggests that the only way to 'see intentions' is to be able to 'understand the environment in which the organism is acting.'[24] Presumably then, in understanding an environment, either a sense of place must at first be achieved or a sense of 'place-referent continuity'[25] or indeed an understanding of the scientific ecology of a place. For Charles,[26] within social affordance we see, within a unique social situation, our 'social partner's minds.'[27] Therefore, if Ryan believes that these situations are social or indeed offer a social affordance of interaction, then he must also believe he sees the minds of elements within the non-human landscape. This is a highly anthropomorphic way of thinking about his relationship to the non-human.

How can we differentiate between anthropocentrism, weak anthropocentrism, or anthropomorphism? And could understanding this shed any light over differing experience within natural landscapes?

> Anthropomorphism describes the tendency to imbue the real or imagined behaviour of nonhuman agents with humanlike characteristics, motivations, intentions, or emotions. […][28]

From what has been said so far within this chapter clearly, some do imbue the *behaviour* of the non-human with humanlike characteristics. This is key, it is not simply that people believe the non-human to be vibrant but that the non-human is anthropomorphised as having both mental and physical characteristics that are akin to humans. Interestingly, Epley et al.[29] use the term 'behaviour': an active, seemingly agentic term.[30] Epley describes this behaviour as 'real or imagined' suggesting that those who anthropomorphise may see these behaviours or they may manifest ideas of these behaviours as humanlike. If we consider behaviour as an intentional noun or verb, 'to behave,' there are connotations of conscious or unconscious activity, both assuming a consciousness.

Helen, Ryan, and Finlay suppose that the non-human behaves in a conscious way. For Epley,[31] the central characteristic of anthropomorphism is a perception of a non-human mind, including conscious experience, metacognition, and intention. Additionally, it also involves attributing humanlike emotions.[32] For Epley, an anthropomorphic person will consider the motivation, intention, and emotions of the non-human, which we have seen. Additionally, Epley states that 'people are more likely to anthropomorphise when anthropocentric knowledge is accessible and applicable.'[33] They do this in order to become effective social agents and if 'lacking a sense of connection to other humans.'[34] For Boyer,[35] however those that do have an anthropomorphic conception of the non-human are extending 'not an 'immediate' or private experience' of 'being a body' but instead the 'mental representations that describe and explain private experience.'[36] Interestingly Bartz, Tchalova, and Fenerci et al.[37] argue that 'individuals who are lonely are more likely to attribute humanlike traits' to the non-human,[38] and they do this in an attempt to 'fulfil unmet needs for belongingness'[39] once again bringing motivation for engagement back to Baumeister and Leary's 'belongingness hypothesis.'[40]

> there's some places where we've got the best of our remnants of the original forest, which is blackwood of Rannoch, you know and some of the Caledonian pine woods or some of the Atlantic oak woods or some of the northern birch woods or some of the temperate mixed woods you know we have got some lovely woodlands[…] and in those very special places, I think you wanna try and keep them very special […] I think it's the right tree in the right place I'm quite relaxed about having other species from other parts of the world, you know growing in the Scottish landscape and being planted in the Scottish landscape and you know, growing well because *they kind of like those conditions* you know they do well under those kind of upland, or windy or wet or dry or whatever the conditions might be and trying to use species which have got some, in some cases, some *commercial use*, the timber is something which we can use and I like the idea of growing our own timber in Scotland […] I'd say in most places outside these kind of *really special places* then you know we should be trying to grow trees for timber as well as growing trees for all sorts of reasons […] (Patrick)

Patrick admires and respects the natural Scottish landscape. Can someone with these feelings towards the non-human, who also manages these landscapes, be deemed anything other than anthropocentric? The management of the natural landscape is the most overt example of placing human value upon the landscape. Patrick, in one breath, mentions what the trees 'like' and how they have 'commercial use.' As someone who not only manages the forested landscape but who advocates the making of products within The Mental Health Initiative as a healing tool, Patrick presents somewhat of a conundrum with regard to these topics. Whilst one minute advocating the maintenance of indigenous species and forestry management allowing to produce firewood and championing the creation of wooden objects, the next he will be addressing the non-human in a highly personified manner. In some cases, he speaks as though the non-human may have its own intention, though rarely its own intrinsic agentic value. He also speaks of the ecological network of living things:

> they certainly interact, there's certainly interactions between individual species and I mean I do think of a woodland as a community you know, and as an ecology. It does include all the non-living as well as the living parts of the system, you know so it's very subtle how all the different bits interact with each other and we don't understand that completely.

For some within the eco-philosophy debate, Patrick would belong steadfast at the opposing end of the spectrum to The Loose Community and those that perceive the non-human as intentional or with intrinsic value. However, if we bring it back to the significance of language, Patrick is saved from this, for some morally corrupt, fate. It is his knowledge, I believe, and the job he *has* to do, that allows him to straddle multiple philosophical standpoints. Perhaps in his personification he maintains an element of moral standing.

Throughout the year, whilst getting to know Patrick on several excursions he said a number of things that seemed to justify the acts of The Woodland Management Group as something other than anthropocentric. Through personification and through giving voice to the non-human, most often to trees with which we worked, Patrick allowed the group a moral high ground in some respects. When tree planting in Fife, he described tree planting as a sacred act: 'Trees are like gold dust [...] You have to hold them carefully and don't bash the bottom roots [it's] a sacred act.' He then describes why he will plant the plum trees away from the apple trees in order that the bees will be able to pollinate them properly—hardly the language of a strong anthropocentrist, though still anthropocentric, nonetheless. Human value is placed in his spiritual reference and desire to grow edible fruit. Patrick believes that woodlands are 'very giving' and that they have many ways in which they 'enhance the quality of our lives,' making his case and viewpoint further complicated. Perhaps we could call him weakly anthropocentric?

In December, on one of Patrick's wood fuel collection days he explained that we were 'working with the tree, showing care, being tough and gentle.' On another excursion in Alyth, I ask him a couple of questions about coppicing and his answer, again, is rather humanising: 'Coppicing is like a coping mechanism for a tree under stress [...] It's weather beaten, so it *makes a decision* to give itself a chance.' The idea that a tree can be under stress is a common use of personification regarding non-human living things. Though common, this still infers that the tree is under emotional or mental pressure. The idea that the tree intends to 'make a decision' and 'give itself a chance' is unarguably anthropomorphic. Yet, Patrick's language intends, rather than to anthropomorphise, to simply communicate, in perhaps the only way possible, that this tree requires the *help* of man to continue to grow. The use of personification aids in justifying behaviours towards the non-human.

Josh, another member of this organisation, who works for most of the time within The Mental Health Initiative, seems to communicate in a similar manner. When Josh talks about felling trees, he is very careful and deliberate. He often personifies the tree, saying things like 'this guy' referring to a tree. He suggests that the tree 'tries to do things,' that they 'just give up,' and that they 'keel over' due to many small damages. As Josh describes the correct way to prune, he says that this means that we will leave a healthier tree who can '*concentrate* on growing up the right way and getting the sun.' Later he tells me this again and he says, taking on the persona of the tree, 'I've been scarred, I've been attacked, I'm packing it in.' This language takes Patrick's and Josh's philosophically complex standpoint and makes it more so. Now they explicitly refer to the mental capacities of a tree. One wonders whether they might do this as a further justification of what they do and of how management decisions are made.

At another excursion in Banff, when Patrick is teaching the group how to fell a tree, he points out the ways in which the other trees that are left behind may benefit from being less crowded: 'This tree says thank you and this tree then says thank you very much.' When I asked Patrick whether he was aware that he did this he told me:

> well it's fun to give trees personalities and they certainly do give us emotional stimulus and I admire trees, you know, I don't identify with trees, I don't consider myself to be a tree [laughter] but I mean, I do like looking at trees and they do make, I do have emotions about trees. The birch tree over there does seem like very delicate foliage and its sometimes called the queen of the forest, so like I can see the femininity of that particular tree as compared to an oak tree which is kind of, seems to me, more masculine you know more kind of rugged if you like.

Clearly Patrick's way of thinking is not anthropomorphic; yet he, as many others we have seen do, personifies the non-human. He acknowledges the effect that trees can have on his emotions but considers this to be down to purely cultural symbolism and gender stereotype. I questioned him further to see if I was able to get to the bottom of his seemingly complex ideas regarding the non-human. I asked him in as straightforward a manner as I could muster: 'Do you consider the non-human as having intention?' His response was somewhat prosaic and full of pragmatism, leading me to conclude that Patrick is rather practical in his considerations of the non-human landscape even if he does acknowledge the benefits of engaging with these natural landscapes. His priorities lie in the maintenance of the landscape:

> well higher animal species, I mean they all have got intention to reproduce themselves, you know to bring up the family, bring up the next generation, to survive, you know to feed, to prey on other species. Animal species have got that kind of volition. When it comes to plants I'm not so sure that plants have any kind of self-awareness or any kind of… I mean they're good at doing what they do I mean, trees are good at growing, they're good

at growing tall you know [laughter] they're good at pumping water out of the soil and up to the branches, up to the leaves, you know, they're extremely efficient organisms which survive and you know have adapted over millions of years, sometimes to become very, very good at what they do, which is survive and reproduce themselves you know but whether… I wouldn't go any further than that in terms of any kind of self-awareness. (Patrick)

If Patrick's comments were taken at face value, he may seem to be nothing more than a conservationist with typically conservationist viewpoints. However, it is important to note that here it is a different kind of empathy towards the non-human than that of The Loose Community. It is one still founded on care however with undertones of the prevalence of man over the administration of this care. He does not completely disregard a reciprocal relationship with nature. In the above quote, he specifies the pleasure achieved when doing this kind of work and in *knowing* that you have helped the environment. He says for example that 'if you're thinning, you're letting more light in so you know you're getting your ground flora so plants are getting a chance to develop, your natural regeneration of trees has got a chance to develop and so on.' Seemingly, a reciprocal care. Aside from a human-centred perspective of knowing that the ecology is helped there is no harm meant.

Patrick believes he is doing *good* in relation to the natural environment, in the same way that The Loose Community believes that they are doing *good* by developing a relationship with the non-human. The facilitators of The Young Adults' Personal Development Project believe that they are doing *good* for the young adults. This *doing good* is of significant interest to this book to decipher why these excursions may be considered as positive to mental wellbeing. We will consider now the case studies who lie at the opposing end of the spectrum with regard to the perception of non-human intention. This time I lay out an analysis of those who value only human efficacy in the bid for personal development, within these excursions. I will then lay out the complexity of believing oneself to be *doing good* within these excursions and how this is linked to the aim of this book—how we might finally understand why these excursions are considered as personally and socially transformative.

The Agency of Self and Group

For most informants, there is not a moment where they explicitly state 'I am an anthropocentrist' or 'I am an anthropomorphist,' or ecocentrist or whichever. When we analyse conversations, comments, and behaviours however, and when we approach these terms by theoretical definition, it may seem that we can associate a term with one person but perhaps not the other. I have dealt with those within my case study groups who could be said to fall on one end of this spectrum: the end that may align itself with ecocentrism, personification, and anthropomorphism. When we move along to the other end, we begin to see groups and individuals who could be deemed anthropocentric in that they have an instrumentalist attitude to natural resources. They utilise the landscape in the sense that they are physically active within, they challenge themselves within, and they are challenged by the physical landscape (its remoteness or terrain). For the most part, these informants do not anthropomorphise the landscape. They utilise the landscape for personal development; however, they are not strongly anthropocentric.

A simple definition offered by Murphy claims agency as an ability to be able to act, do, and effect one's own attitudes, behaviours, and thus experience:

> By agency I mean the ability of individuals to act in the world to effect change, intentionally and unintentionally and with foreseen and unforeseen consequences.[41]

Agency is a control over what one does and what one feels afterwards. This is a sense of agency over experience that is not affected by the external power of a non-human other. This is a personal agency over the experience of oneself within the group and landscape—to act, and to challenge oneself, to be autonomous over one's experience.

> The capacity to exercise control over one's own thought processes, motivation, and action is a distinctively human characteristic. Because judgments and actions are partly self-determined, people can effect change in themselves and their situations through their own efforts [...][42]

Self-efficacy is an ability in oneself to achieve a desired goal, be that an ability to manage one's anger, enable oneself to cope in new situations, drop a harmful habit, or prepare oneself for a job interview. For The Young Adults, The Mental Health Initiative, and The Community Living Initiative this is the generic standpoint regarding their experiences in natural landscapes. They have individual agency, and they themselves carry out actions and interventions that produce a particular effect. They have the ability to act. As sole actors, they shape their experiences and thus, arguably, the outcomes of engagement. They have moral agency in that they choose, with intention, within a given circumstance, to behave and think in certain ways. Whereby the previous cases perceived the non-human as having moral agency or intention, these groups, for the most part, perceive that it is their own human agency that brings about effects within their own experience. It is their own intentions, behaviours, and thought processes that ultimately form the experiences that they have within the natural landscape and with the group. These are social experiences; however, their social networks are more limited, to the human group, within the minds of these informants.

Bandura[43] discusses agency as the capacity in which humans may exercise control over their thought process, motivation, and action. He believes, unlike my previous informants, that this is a distinctly human characteristic. For Bandura, this agency is what enables people to elicit control over themselves and their own situations through effort and action. In a more simplistic way, with reference to the groups I will now discuss, it is our own self, and fellow humans, alongside the utilisation of the landscape that will affect any kind of transformation.

Bandura[44] tellingly associates an optimistic sense of personal efficacy with human attainment and wellbeing. Additionally, he states that those who already judge themselves as 'highly efficacious' will expect positive outcomes and those who assume bad results or personal performances will 'conjure up negative outcomes.'[45] This would then suggest that within my case study groups, all individuals began with a positive sense of their own abilities. In my experience, this is not the case. In most cases, this positive sense of personal efficacy was often an outcome of the shared experience—though definitely a desirable outcome on the part of the group facilitators. Bandura seems to miss the potential instances for those with difficult life circumstances, learning difficulties, or mental health issues, or where organisations are aiming to alter negative perceptions of self-efficacy and ability. Self-efficacy and anxiety, or belief in oneself affects perception of risk[46]; therefore, a person becomes willing to 'go for things.' These journeys represent different things for those with or without self-belief to achieve what they set out to do.

For Holzkamp,[47] the human capacity for action comes about due to the particular social character of human life. He focuses on the fact that individuals strive to 'extend their influence over their living conditions, an effort that necessitates participation in societal action.'[48] For Bruner,[49] human action is 'under the sway of intentional states, such as beliefs, desires, emotions, and moral commitments, states which in turn are interwoven with culture, society, and history.'[50] This social network does not include the non-human in an agentic way, but it does by way of meaning making. For Holzcamp, meaning is a 'possibility for action'; these meanings are relational, societal, historical, and not 'deterministic triggers of action' meaning 'they are not physical, chemical, or biological or constraints, but indicate or signal a range of options, of possibilities for action.'[51]

Self-Development

Culture deciphers the way in which case study group individuals project meaning on to their surroundings. This culture, for the young adults, is one in which their personal narrative has very little to do with the natural world. Instead, it is very much to do with nature as an unfamiliar other. The culture in which they live, at least if we take The Young Adults' Personal Development Project as a microculture for these young people, is based on one's *ability* to help oneself. This has an effect on the thoughts of the young adults and the group experience:

> I think self-development is key, I think that's kind of a central theme that we offer and try to inspire or instigate within them and I think part of what they're looking for is not achievable where they are when they come to the course so it requires them to develop—to open to kind of develop different

> sides of themselves in order to kind of, you know, fulfil those needs. I think the course… hopefully the course enables that for them or creates the context for them to do that but it's very much driven by them, we can't do that, we can only provide the context I think and you know kind of prod them and yeah, it's up to them. (Rory)

Rory speaks of *self*-development and not personal development. To me, this suggests that, for Rory, the programme is about developing the self as opposed to doing something that will simply further personal or pragmatic goals though self-development may aid in this. His use of the word 'self' suggests that this is very much an internal development and one that relies on not only a sense of self but on one's own agency. He confirms this as the central theme, at least in his view. This is an important one as he is the primary facilitator and contact point for the young adults. Rory's job is to 'inspire.' He does not explicitly state that this is the only thing within these excursions that may inspire, but that it is a human social relationship that will do so.

For Rory, 'part of what they're looking for' (again not explicitly stated) cannot be achieved or attained 'where they are when they come to the course.' By this I believe he is not only referring to their physical position but their 'head space.' When speaking of achievement, Rory inadvertently hints towards efficacy and agency as defined above, an ability to bring about desired results and to act. Prior to attending the course then, he believes that the young adults were lacking in efficacy and agency due to their circumstances. The course is meant to change this situation. It is believed to allow for a sense of positive efficacy and action, based on the self and social relationships built. For this to happen, the young adults are required to be open to 'different sides of themselves' to 'fulfil those [not-articulated] needs' (Rory). The young adults are tapping into 'different' aspects of themselves in that each individual crosses several metaphorical liminal thresholds in the mind—be that through uncertainty, challenge, or associating experience with one's personal narrative. The course, rather than only the outdoors space, provides the context for this and 'it' is then, according to Rory, driven forward by the individual. 'It' is also not quite specified. Presumably though 'it' refers to self-agency and efficacy to positively influence one's own life choices.

> […] your sticking at it and doing like, *to show you can do it* and its, its good just. I never thought in a million years I'd be able to like walk out over hills and that, never, no, it's good…
>
> I felt it was a lot different than what it is staying at home and doing stuff, as in you've got everything to go by. You've got a microwave and everything but here you've got to use stuff to you [sic] advantage. If you want boiling water you've got to use the stove which is a lot. It's a good experience. It was good to put everything aside. To put electric and everything at your side and just focus on stuff that… *it just shows that you can do stuff without having the tools* to do stuff what you normally do [sic] if you know what I mean. There's more to life than sitting in the house doing nothing.
>
> […] I understood that there was no like electricity or anything, there was no [mobile phone] service or anything you can … you're out in the wilderness, you can't, you can't, you know what I mean? You know what I mean. It's all about survival you can't just use your phone and OK Google this, that and this. (Jonathan, *emphasis my own*)

Jonathan's perspective is relatively rare, as someone who suffers with not only mental health issues, but autism and further learning difficulties, his experience is relatively unique. He is in fact one of three within The Young Adults' Personal Development Project who is on the autistic spectrum. In his own words, he feels he is highly functioning and does not consider himself on the same 'level' as the other men in the group. Jonathan's personal difficulties stem from social difficulties and difficulties in engaging in what for some might be relatively basic. We must acknowledge then that this experience was always intended to be one that would challenge him in particularly social ways with his peers. Jonathan's participation in the project came at the beginning of his 'independent travelling,' something he had little experience of. From the outset, personal challenges, perhaps basic or fundamental social skills, were at the fore. The project was very explicitly a set of challenges and obstacles to be overcome for Jonathan.

The concept of 'sticking at it' heavily implicates one's own determination or indeed intention to complete a task. Though his own personal challenges and emotions about this are often skirted around, there is a strong sense of discovering one's own personal efficacy in Jonathan's comments. He states, 'To show you can do it.' I wonder though who he wishes to *show* this to—himself or some other? Either way, we see his sense of self implicated again—either his ideal sense of self or his ought sense of self. He speaks of his previous lack of self-belief: He 'never thought in a million years [he'd] be able to' do the things that were expected of him. In this instance, these tasks are physical expectations, but often I also saw within Jonathan a real attempt to control emotions that had in the past dictated social circumstance—for example, his anger, his lashing out at others, both physically and verbally and his regular bouts of petulance and seeming fight or flight responses to social challenges.

Jonathan talks of 'us[ing] stuff to [his] advantage.' Clearly in Jonathan's mind the use of material objects, perhaps the landscape too, is about using them to benefit oneself either practically or mentally. This is one of few experiences for Jonathan that required him to be able to complete tasks without having the modern conveniences that he has back at home in Haddington. Interestingly, this is Jonathan's most relevant comparison—the lack of presence of mod-cons and technology. He refers rarely to the landscape, instead focusing on what is not there. It is what was *not there* that provided most of Jonathan's challenges, at least those that he chose to speak. He also refers to his overreliance on the internet: 'It's about survival [...] [you] can't just Google.' For me, this is a reference to using one's own thought process and learning rather than relying on a search engine, and Jonathan's sense of survival comes down to this—how might he cope without the modern conveniences that dominate his everyday life? (We saw earlier that Jonathan's social communication was almost entirely web based). For Jonathan, then, it is about deprivation and what is *not* available rather than what is, that is the landscape. For Jonathan, he has his sense of self-efficacy ('you can't rely on anything but yourself') and the challenge of coping *without* what he is used to. For Jonathan, any sense of personal development is about being outside of his urban comfort zone. He makes another comparison to his *everyday* experience. He suggests that there is 'more to life than sitting in the house'; this was his everyday reality. Aside from Jonathan's reference to wilderness, the overtone of his comments is that of a reliance on oneself, further, reliance on himself to overcome personal challenges whether physical or mental, social or cultural.

'Designed' Excursions

Regardless of Rory's belief in the prevalence of self-efficacy as a determinant of positive transformation within these excursions, he does not entirely negate the pertinence of location. It is not purely self-efficacy that will bring about change within the young adults. It is after all deliberately an outdoor course and not one centred within a purely urban environment:

> We are an outdoor course, it's very much an outdoor course. [...] I'd say it's symbiotic really with... You know what our intentions are and through experience and design I suppose that we know that we can create the context for those developments in young people by going to these places and doing these activities [...] it's not necessarily, it's by *design* from our sense so is intentional but only because we know these places have the ability to change people. If that makes sense? And, I think that for me personally that's born from personal experience. Because *I know that mountains and rivers and climbs and what not have changed me* and I have learnt and grown through them and had my own personal development experiences through them. So, it's kind of like in that sense it's kind of like a lineage isn't it? Like handing down something through a generation maybe. (Rory)

The Young Adults' excursions are, in Rory's own words, designed. The facilitators and the organisation create a context in which the young adults can develop. They do not rely on 'seeing what comes,' in the way that The Loose Community do. Instead, they rehearse excursions, theory, and predetermined agenda for personal development. There seems to be considerably more control over the types of experience had, at the very least, the types of experience spoken about. There seems little space for discussion surrounding the anxieties felt by the group or the more phenomenological, psychological experiences had, whether in relation to the self or the non-human. These excursions are somewhat formulaic. Though of course individual difficulties and issues, as well as group dynamics require a degree of flexibility in activities, the basis of the excursions are often unchanged from group to group. In a sense, facilitators *know* what these spaces *do*, they take them to these spaces, and they have well-rehearsed experiences. This is particularly evident in the organisation's reliance on Tuckman's model of group development. It is presupposed that the group will proceed through stages from forming to storming to norming and so on. We have seen though that this is certainly not always an applicable or accurate process. It is also not necessarily representative of true experience, instead focused on the desired developments that will enable the young adults to go on and function within our society—in a 'better' way than they do currently.

Rory tells me that '[they] know these places have the ability to change people.' This *knowledge* cannot but be based on the facilitators' own subjective experience, personal narrative, and context, as well as previous experiences of previous groups. He describes it as 'a lineage' that is based on a passing on of personal beliefs and direct influence on how the landscape is approached and thus experienced. To me, there seems to be something missing in Rory's comments. He does not provide an explanation as to why these locations have effect and why they know that these landscapes are useful in self-development, only that this is how he experiences them, and this is the foundation on which the project is based. It seems to be about self-efficacy and it is believed to work—exactly why is left to a cultural knowledge, seemingly held by the facilitators.

> [experience is carried down in] almost like in a shamanistic [sic] kind way. You're going to go to these special places and you connect with them. And allow them to kind of connect with you. And that is by design but equally it is not so. (Rory)

The above comment, though acknowledging the idea of a shamanic relationship to the landscape and an elder to youth passing down of knowledge, still chimes of a formulaic approach: 'You're going to go [...] you connect [...]' and they will 'connect with you.' It is planned in such a way, or 'designed' in such a way that there is very little scope for anything other than obligatory experience. He speaks of these locations as 'special places' again leaving it rather phenomenological with little explanation as to why they are special, or why indeed they warrant being thought of in such a way. Returning to the idea that these young people have had little, if any, experiences of these types of landscapes that would allow them to *know* the landscapes in this way, we are left with an enforced cultural manifestation and general knowledge of what may be on offer. Until experiences are embedded within one's own personal narrative, this will most likely remain so.

Is Landscape Relevant?

Within The Young Adults' Personal Development Project, Rory's point of view is not necessarily representative of all the facilitators. Anna is another of the group's leaders who had significant contact and provided a support to the young adults. Her perspective is important to address, particularly if we are interested in the influence of perceived authority roles on the experience of the young adults.

> [...] I think it's the relationships that are probably the most important thing [...] but building on those relationships in the natural environment I think is really important and the ability to do that, I get the impression that's different being out in the bothy in Glen Etive than could be recreated, you know, in a community centre here [...] I kind of think that the environment's really important to it and allowing them the sort of space and time away from the sort of pressures [...] and the kind of anxiety driven, kind of, environment that sometimes the city can be for people. So, I feel like, actually getting that feeling, that feeling of perspective as you've just said—of being, you know, were just, little people in the vastness that is the world or the universe or however you wanna look at it but you can't get that perspective, or I would struggle to see the group getting that perspective from in here. (Anna)

Though Anna, like Rory, believes that it is the human social relationships that are key to the 'success' of these excursions, she also acknowledges the space as a significant factor. This could, at first glance, be understood as a rather ecocentric positioning. However, when she specifies Glen Etive, she does not refer to Glen Etive's specific landscape characteristics. Anna only states that the bothy is there (with that, the suggestion of limited mod-cons). She also comments on the space and time away from external and internal pressures that the excursions to this location provide. The space and the distance from the urban landscape is not particularly distinguishable from another outdoor, wide open, landscape. This, to me, suggests that these excursions are considered successful due to a weakly anthropocentric perspective. The space is utilised independent of any intrinsic value that it may have. The landscape's value is based on its scope for providing space, distance, and perspective to participants in a manner suitable to the organisation's agenda.

> I mean we've got a great big hill in the middle of our city here,[52] there is no reason why, if you're feeling really stressed and anxious you can't take, you can't walk out. As, well it depends where you live in the city but it can take you 10 minutes […] [it's] beneficial to different people but that actually, I feel like […] you know part of what we do is say, that this is a good coping strategy. If you're feeling really anxious and brought down by things that are happening in your life and, then this might help. (Anna)

For Anna, granted due to the unusual landscape of Edinburgh's town centre, 'getting out' does not necessarily require the young adults to leave Edinburgh. Presumably then, an *imagined* distance is equally powerful. An imagined distance, I suggest is a feeling of being distant from one's everyday life, as one might be atop Arthur's Seat, yet perhaps only a mile away from home. Being able to go outdoors, and to engage with the natural landscape is a useful coping strategy for Anna—one that can help to combat stress and anxiety, regardless of its exact location and characteristics. The space is somewhat irrelevant, so long as it is removed from the everyday.

The perspective of being a small element within a vast world, in Anna's opinion is another important sense within these experiences to consider. Anna hopes for the young adults to gain some perspective of where they place themselves within this world. This potentially places Anna in a different position along the spectrum to others and might suggest that she considers a wider relative non-human network in which we may place ourselves. For Rory, this network, in which he includes nature, is 'an artifice,' as is any difference between the city and Glen Etive itself:

> It goes two ways and it's like, just because you go and sit in a, in a glen and like [say] 'ah this is beautiful,' doesn't necessarily mean that you are opening yourself up to the universe in a sense but you [...] it might be that if you were here and something clicks for you then you start opening up and that could be equally powerful experience.

Rory's comments echo my belief that after building on one's experience and thus personal narrative and ability to abstract ideas, or in Rory's words, 'when something clicks' or one 'starts opening up,' then the aesthetic experience of the landscape may add significance to these experiences. Until that time though, it seems The Young Adults' Personal Development Project places the human, with their own efficacy and intention, at the helm of these experiences. Further indication that The Young Adults' Personal Development Project comes at these excursions with a particularly weakly anthropocentric philosophy is the main claim that, quite simply as Anna states: 'It can be really good for you,' to get out into natural space.

Notes

1. Attfield (2011: 41).
2. Attfield (2011).
3. Rautio (2013).
4. Rautio (2013: 447).
5. Rautio (2013: 451).
6. Rautio (2013).
7. In Attfield (2011).
8. I had mentioned that when I thought of a sycamore tree I couldn't help but think of the song 'Dream a Little Dream of Me' by the Mamas and the Papas (1968) with the lyric 'birds singing in the sycamore tree, dream a little dream of me...'—a rather romantic culturally defined notion with reference to a sycamore tree.
9. Epley et al. (2007: 867).
10. Boyer (1996: 83).
11. Bennet (2010).
12. Bennet (2010: 05).

13. Cloke and Jones (2003: 199–200).
14. Cocks and Simpson (2015).
15. Cocks and Simpson (2015: 220).
16. Attfield (2011: 30).
17. Attfield (2011).
18. In Attfield (2011).
19. in Cocks and Simpson (2015: 218).
20. Cloke and Jones (2003).
21. Cloke and Jones (2003: 200).
22. in Boyer (1996: 87).
23. Charles (2011: 137).
24. Charles (2011: 138).
25. A sense of a familiar landscape, for example being in a forest when one has found a sense of place in another forest, one might also feel a sense of familiarity or *place* in the said new forest (Csikszentmihalyi and Rochberg-Halton 2001).
26. Charles (2011: 149).
27. Charles (2011: 149).
28. Epley et al. (2007: 864).
29. Epley et al. (2007).
30. Though the Oxford dictionary (2017) defines behaviour as 'the way in which one acts or conducts oneself, especially towards others' and 'the way in which an animal or person behaves in response to a particular situation or stimulus,' it also gives us this definition: 'The way in which a machine or natural phenomenon works or functions' deeming behaviour as something that could be said of a human, a car, or presumably a tree or the like.
31. Referencing Gray, Gray and Wegner in Epley et al. (2007).
32. Epley et al. (2007: 865).
33. Epley et al. (2007: 864).
34. Epley et al. (2007).
35. Boyer (1996).
36. Boyer (1996: 87–88).
37. Bartz et al. (2016).
38. Bartz et al. (2016: 1644).
39. Bartz et al. (2016: 1644).
40. Baumeister and Leary (1995).
41. Murphy (1992: 312).

42. Bandura (1989: 1175).
43. Bandura (1989).
44. Bandura (1989: 1176).
45. Bandura (1989).
46. Bandura (1989: 1177).
47. in Brockmeier (2009).
48. in Brockmeier (2009: 219).
49. in Brockmeier (2009).
50. In Brockmeier (2009).
51. in Brockmeier (2009: 222).
52. Arthur's Seat in Edinburgh.

References

Attfield, R. (2011). Beyond Anthropocentrism. *The Royal Institute of Philosophy Supplement, 69*, 29–46. https://doi.org/10.1017/S1358246111000191.

Bandura, A. (1989). Human Agency in Social Cognitive Theory. *American Psychologist, 44*(9), 1175–1184. American Psychologist Association, Inc.

Bartz, J. A., Tchalova, K., & Fenerci, C. (2016). Research Report: Reminders of Social Connection Can Attenuate Anthropomorphism: A Replication and Extension of Epley, Akalis, Waytz, and Cacioppo (2008). *Psychological Science, 27*(12), 1644–1650. https://doi.org/10.1177/0956797616668510.

Baumeister, R. F., & Leary, M. R. (1995). The Need to Belong: Desire for Interpersonal Attachments as a Fundamental Human Motivation. *Psychological Bulletin, 117*(3), 497–529.

Bennet, J. (2010). *Vibrant Matter: A Political Ecology of Things*. Durham/London: Duke University Press.

Boyer, P. (1996, March). What Makes Anthropomorphism Natural: Intuitive Ontology and Cultural Representations. *The Royal Anthropological Institute, 2*(1), 83–97. http://www.jstor.org/stable/3034634

Brockmeier, J. (2009). Reaching for Meaning: Human Agency and the Narrative Imagination. *Theory and Psychology, 19*(2), 213–233. Sage Publications.

Charles, E. P. (2011). Ecological Psychology and Social Psychology: It Is Holt, or Nothing! *Integrated Psychological Behaviour, 45*, 132–153. https://doi.org/10.1007/s12124-010-9125-8.

Cloke, P., & Jones, O. (2003). Grounding Ethical Mindfulness for/in Nature: Trees in Their Places. *Ethics, Place and Environment, 6*(3), 195–213. https://doi.org/10.1080/1366879042000200660.

Cocks, S., & Simpson, S. (2015). Anthropocentric and Ecocentric: An Application of Environmental Philosophy to Outdoor Recreation and Environmental Education. *The Journal of Experimental Education, 38*(3), 216–227. https://doi.org/10.1177/1053825915571750.

Csikszentmihalyi and Rochberg-Halton. (2001). The Meaning of Things: Domestic Symbols and the Self. *Contemporary Sociology, 12*(4). https://doi.org/10.2307/2067526.

Epley, N., Waytz, A., & Cacioppo, J. T. (2007). On Seeing Human: A Three-Factor Theory of Anthropomorphism. *Psychological Review, 114*(4), 864–886. The American Psychologist Association. https://doi.org/10.1037/0033-295X.114.4.864.

Murphy, P. D. (1992). Rethinking the Relations of Nature, Culture and Agency. *Environmental Value, 1*(4), 311–320. White Horse Press. http://www.jstor.org/stable/30301328

Rautio, P. (2013). Being Nature: Interspecies Articulation as a Species-Specific Practice of Relating to the Environment. *Environmental Education Research, 19*(4), 445–457. https://doi.org/10.1080/13504622.2012.700698.

The Mamas and the Papas. (1968). Dream a Little Dream of Me. In *The Papas and The Mamas* [CD]. Dunhill.

The Oxford English Dictionary Online. (2017). (Behaviour). Available at https://en.oxforddictionaries.com/definition/behaviour. Accessed 30 Aug 2017.

7

Conclusion: Performed Identities and Being a Good Person

The diagram below shows first, at the top of the diagram, what we understand quite universally—that society and culture may have intention (this is shown by the small light bulb icon above). This follows down—a group can have intention and of course an individual may have intention. If you look on the left however, you will see that for some, the living-non-human equation can also be endowed with an intention (for some explicitly and for others metaphorically). This figure demonstrates a fork in the road. This is where individuals as part of a group consider the landscape in one way or another. This can be due to societal and cultural influences, intrinsic beliefs, or indeed due to the agenda of the group or organisation. This diagram poses the question: what encourages people down one road and not the other? We will consider that it is in wanting to not only belong, but to self-verify—specifically by feeling that one is doing *good* (Fig. 7.1).

I was with The Loose Community, and we were packing up for the day after our day of mindfulness and Goethean observation. I knelt on the floor, packing my note books into my bag, when someone asked what I was 'up to' in the morning. I hesitated, anticipating that my answer would most likely not be palatable to this group. I gulped, laughed awkwardly, and informed them that I was going on a wood fuel collection day with

R. Crowther, *Wellbeing and Self-Transformation in Natural Landscapes*,
https://doi.org/10.1007/978-3-319-97673-0_7

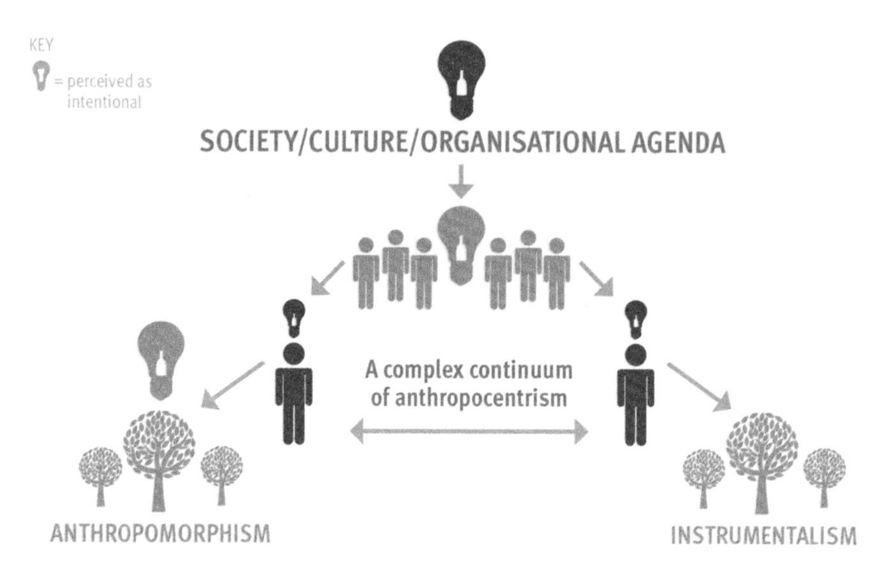

Fig. 7.1 A complex continuum

another case study group, The Woodland Weekenders. I would in fact be chopping down, predominantly due to its thriving in Scotland … sycamore trees. The silence for a moment was unbearable before Christie exclaimed, 'Don't worry we know you're a good girl […] be strong,' at once allowing me the sense that I was not only perceived as part of this group but also that they believed I held the same philosophical standpoint towards the non-human. Whether I do or don't is quite beside the point here. This comment also alerted me to the fact that they believed the way they interacted with nature was the correct and *good* way to interact. The inference being that what The Woodland Weekenders do, despite their altruism and aims of helping those with mental health issues, is *bad*. I know of the positive intentions of this group; however, I am also aware that many people would feel morally uncomfortable chopping down a living tree.

The complexity of this was intriguing. If all my case study groups were aiming for some form of bettering, either in themselves, for others, or for the environment—whether through aiming to understand the intrinsic value of the non-human, aiming to conserve and sustainably manage the landscape, or aiming to better the lives of underprivileged youths—all relationships with nature were aiming to do *good*. If all achieve *good*, then how can case study groups and activities be perceived as so very different between themselves? The positive aspects of each of the case study groups approaches and outcomes, often complex and undefinable, are difficult to dispute if we are talking of the aims of these groups—to improve well-being. There is a sense though, within the remit of outdoor engagement, of a certain vying for the moral high ground.

What is the correct way to be in nature, and does it make any difference to experience if we are thinking about positive personal transformation?

This was an exemplary moment within fieldwork. It highlighted a stark difference between one group and the other. Christie's comment 'we know you're a good girl' proved helpful in understanding complex relationships with the non-human in these instances. This conversation acted as a moment of revelation within my fieldwork, highlighting something that may underlie all claims to positive self-transformation. It concerns our moral positioning towards the non-human and living natural landscape and what it means to be a *good person*. It brings us back to ideas relating to self-verification and notions of the ideal self—what Cocks and Simpson[1] refer to as 'satisfaction with oneself.'[2] One might argue that, for the most part, people want to be *good*—they want to associate themselves with acts and others who do *good* or who are considered favourably, particularly if we are speaking in the remit of positive self-transformation. So, it would follow that the reason for participating in excursions into the natural landscape would be concerned with becoming or reaffirming oneself as a *good* person (Fig. 7.2).

There is a lack of self-awareness with regard to the complexity of what it is to be anthropocentric in one's relationship towards the natural landscape within my case study groups. There was also a lack of awareness with regard to the hypocrisy of deeming one form of relation over another as *good*. So, I asked myself—does it matter? If each interaction with the natural landscape could be said to be, in its own way, personally positively transformative, then surely each excursion is doing social good. The primary aim of my case study groups therefore is negating the importance of engaging in any form of moral debate regarding hypothetically 'correct' interactions with nature—particularly if we are interested in human wellbeing. Does one form of engagement do *more good* than

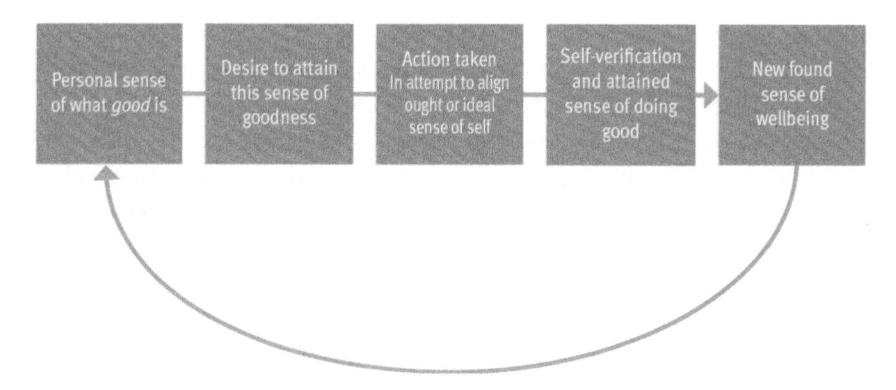

Fig. 7.2 Being a good person

another? Do those that anthropomorphise experience higher wellbeing? I could not say. Determining the value of wellbeing or what constitutes a better sense of wellbeing is so highly subjective and incomparable. I engage with this theoretically to help in concluding this book.

I take Hill Jnr,[3] philosopher and ethicist, as a starting point:

> Barring special explanation, we can expect that virtuous persons will value nature for its own sake – at least they will not regard the natural environment merely as a means to human welfare or as something whose treatment is constrained only by human rights [...][4]

Virtue, meaning goodness, morality, righteousness, decency, purity, nobility of soul and spirit, integrity, and so on alongside the above quote leaves many informants in a morally precarious position! If we were to heed Hill Jnr, perhaps all of my case study informants would be lacking in virtue. After all, as I have shown none can truly relate to nature with only an understanding of its intrinsic value as that value is inaccessible to the human mind and body. To value nature for purely its own sake as I have discussed is a rarity within my experience working within the field. As Val Plumwood[5] concedes:

> Humans might wish to extend moral respect to nature, or recognise that parts of nature are a little like themselves and so have a moral right to life and well-being, but the parameters for inclusion remain human.[6]

Hill Jnr[7] argues that these virtuous persons would at least 'not regard the natural environment as a means to human welfare.' Whilst none of the case study activity overtly pillages the landscape nor do they excessively use for human consumption, The Woodland Weekenders do manage the landscape. They also gather wood fuel. The Mental Health Initiative manufactures objects out of wood, and all case study groups instrumentalise the landscape for human welfare in the sense that they aim for personal and social positive transformation. I'm sure many would agree that these informants do not deserve such a smear of character; however, Christie explicitly tells me that regardless of my actions, she knows I am a '*good* person,' and the implication that others are *bad* in comparison still lies.

> […] valuing of natural environments is essential to (and not just a natural basis for) a broader human virtue that we might call 'appreciation of the good' […] The virtue that I have in mind now, broadly speaking, is a manifest readiness to appreciate the good in all sorts of things, and not just as an instrument or resource for something else […] arguably it is widely (and rightly) recognised as a human virtue or excellence, an admirable trait of character.[8]

To value natural environments is for Hill Jnr[9] associated with a broader human virtue: 'the appreciation of the good'—within my case study groups each individual within the group appreciates nature in their own way. Some appreciate that they walk in nature. Some appreciate that they create things from materials found in nature. Some appreciate the challenge of being in nature as a welcome diversion from the norm, and some appreciate nature as a soulful and communicative network of beings. All though do appreciate nature, whether intrinsically, extrinsically, as part of one's narrative, or through the agenda of a group, they appreciate that there is something inherently *good* in being within natural spaces. For the most part, then, we could say that all are virtuous as all appreciate that there is good in these natural landscapes. All could be said, as virtuous people, to be *good* people then. For some informants, they do not believe this to be quite the case. All may appreciate, yet not all do so without appreciating the non-human as an instrument to another thing. At the very least we could say that all within these case studies consider the non-human landscape as an instrument, in that they appreciate it as an instrument to achieving another thing—this other thing being personal transformation and a sense of belonging, self-verification, or indeed wellbeing. There goes everyone's virtue then! When all are 'perspectival anthropocentrists,'[10] all cannot but utilise the landscape even in enchantment and anthropomorphism.

> The flourishing of nature is constitutive of human wellbeing.[11]

Aristotelian ethics may offer a more fitting, or certainly a less bleak picture of what these groups attempt to do in their engagement with nature. Aristotelian flourishing, or the flourishing of all being dependent on the flourishing of man,[12] allows us to consider things differently, perhaps differently to the current trends within research in the Anthropocene—one that may allow for a compassionate view of man and human wellbeing in relation to the natural landscape through supposed anthropocentrism.

All the case study groups discussed are interested in the 'flourishing of man' in that they are interested in the wellbeing and personal, positive transformation of the group individuals regardless of the agenda behind this. They may be considered as anthropocentric, but 'not objectionably so.'[13] The examples of care, albeit anthropocentric as shown by The Mental Health Initiative and The Woodland Weekend Group, as well as the imaginative empathy of The Loose Community towards the natural non-human can be separated from the human-value-based activity of The Youth Development Project. These case study groups, whilst providing opportunity for the 'flourishing of man,' or their own positive transformation, subsequently provide a reciprocal care in the flourishing of the landscape. Some being more hands on with regards to this care. The difference between these case study groups and The Youth Development Project perhaps can be brought back to abilities of abstraction among the informants. The young adult's activity is based on human social interaction within the landscapes, the landscape becoming rather a backdrop—perhaps the most instrumental of case study groups then. However, separate to interaction with the natural there are also instances of only human, social interaction, another activity that aids in human flourishing.

> While a reflective awareness of nature's otherness can provide the benefits already mentioned, a comparative unawareness of it (as in failure to reflect on the world of nature at all) need not spell lack of perspective or absence of a sense of proportion. For these benefits, could be derived from other sources, such as human conversation [...] Such unawareness certainly need not betoken egoism, from which we may be rescued [...] Hence, while some basic awareness of nature's otherness may be a precondition of human life and thus human well-being, explicit or reflective awareness of this otherness cannot be regarded as constitutive of or essential to human flourishing, even though it can importantly contribute to such a life.[14]

The young adult's knowledge of how to engage within these environments is determined by the design of the excursions by facilitators with the agenda of *doing good* by 'the young people,' rather than determined by an understanding of their own efficacy to do *good*, be *good*, or see *good*. This I believe is down to the lack of association of nature with their own personal narrative but also of a lack of association with being a *good* person in the minds of the young people within their own personal narrative. Many of the young adults articulate feelings of self-resentment, self-loathing, lack of confidence, and self-belief, as well as chastise themselves for issues that they may have. To presume any of them without virtue (as Hill Jnr might!) would be catastrophic. Instead, what the organisation does is provide the young adults with contexts in order to allow them to potentially appreciate the *good* within these circumstances and other people and thus feel virtuous in their actions—to consider themselves as *good people*.

All case study groups believe that they, in respect of personal transformation and positive self-verification are doing *good*. Only by defining good, could any true differentiation be made between the groups in relation to perceived personal transformation through experience.

The Bettering of Human Experience

The diagram below demonstrates the complex attitudes and philosophies that may be adopted by an individual in relation to their emotional attachment, relationship, and behaviour with and within natural space.

Often a person begins with a fundamental understanding (whether in as many words or not) they perceive the non-human as intentional, whether they are unaware of this possibility, or whether they consider only human agency as an option of viable mediator over experience. This fundamental understanding is what leads a person (with the motivation of finding a sense of belonging and self-verification and feeling one is doing good) to partake in activities and excursions with any given group. The personal philosophical perspectives that are then adopted within a space is complex but most certainly influenced by these motivations and the agenda of the group, the group dynamics, and intrinsic values. There are many more intangible elements (Fig. 7.3).

It is within the crossing of the metaphorical liminal sites of self, group, and the way that the non-human natural landscape is perceived that an understanding of the shared experience of natural landscape can be found.

I will begin to conclude with a poignant moment of identity performance from within the field. In doing this, I bring the book full circle to my discussion of social drama, performance, and transdisciplinarity at the beginning of the book. With this, I will pose further questions which are opened beyond this day within the field to further threads for probable future research to follow this text.

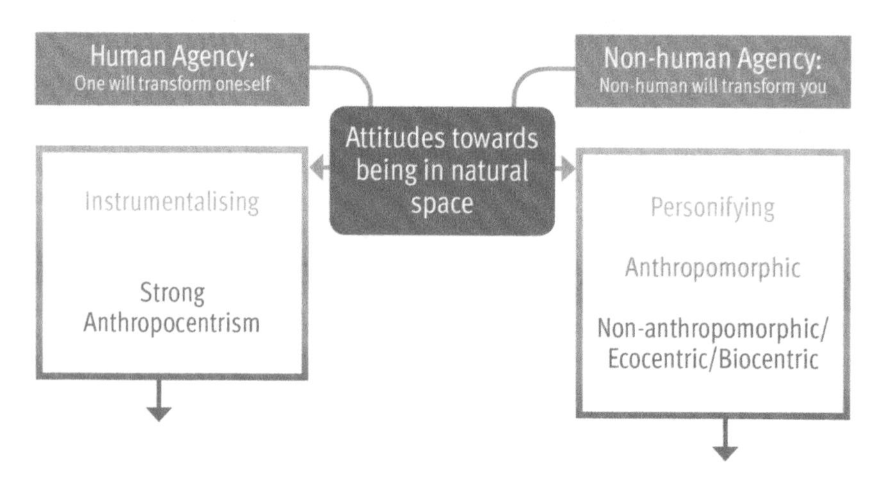

Fig. 7.3 Complex anthropocentric attitudes towards natural landscape amongst groups

Returning to the Earth: A Final Performance

Rodica[15] was a member of the Loose Community. I first heard of Rodica's death over the phone. She had been driving down the A68 near Dalkeith, a busy motorway, and was killed instantly in a head-on collision. Rodica's husband had asked Helen to pass on the devastating news to the group.

Rodica was a Romanian; she had been living in Scotland for over a decade. She was an interpreter, a translator, a storyteller, and a therapist, and she studied shamanic healing, delivering sessions with her husband in Edinburgh. I knew Rodica for only a few months; we had been on several weekends together, and she had once lent me boot socks when my feet were soaked through after a walk in the snow. She had driven me home from a weekend in Callander too, where we stopped by Loch Lomond for a coffee. Myself, her husband, and Rodica stood gazing, exhausted, out over the loch. Her death struck me hard. I knew, due to her involvement in The Loose Community that she was 'working through things' to do with her sense of self, her relationship to the world, and her practice. She was killed suddenly, so unexpectedly doing something that felt, to me, so removed from the natural environments in which we met. It is Rodica's burial that will help me to bring some of this book's key themes together.

I was collected by Helen and Catherine on a dreich day. They were picking me up to make our way to the site of the funeral. I sat quietly and awkwardly in the back of the car. Emotions were high, and it felt as though people were unsure of what to talk about. We exchanged a few words, perhaps, like me, they contemplated Rodica and how on earth we had ended up here.

The burial ground was a dense mature woodland at the top of a winding lane, lined with trees. It was August, the sun flitted through the leaves and warmed our faces. There was, within the sadness, a sense of pulling together. A leaf, fallen from a tree landed on my face to Helen's delight. It's lucky, she said, so I duly placed it in my handbag. The surroundings were familiar; yet the circumstance in which we met that day, so entirely different. The group dynamics was gentle, kind, still open; and supportive but with a tangible heartache amongst us. We were in Binning Wood, a memorial woodland that facilitates natural burials, in Tyninghame, East Lothian, close by to where Rodica and her husband lived together. The site is intended for those who wish to 'minimise the impact of their passing.'[16] Though the concept of 'natural burial' can mean so much more to individuals and families, and in this case, this was so.

We saw a movement of people and slowly made our way to the lane of trees where we gathered with the rest of Rodica's friends and non-immediate family. We stood beneath the trees and lined the passage way for the hearse. We stood, woodland at our backs waiting, watching. The hearse appeared. I, and others winced at the sight of Rodica's coffin; a wicker basket, a clear signifier as to who Rodica was.

It felt like a long walk behind the hearse, with a group who I had walked with many times before. It was muddy and damp under foot. We stumbled down a slight slope into a woodland space, bathed in light, open. We stood together, tightly, in a line huddled as we watched the family gather around the coffin. It was lowered into the ground by the couple's brothers and her husband. I couldn't help but cry silently throughout the burial and gaze around at the people, the trees, and the hole in the earth. Before we left, we were invited to kneel and toss flowers from a basket on to her grave. Helen saw the 'leaves, seeds, beasties' beneath us and 'imagined scooping it all up in [her] hands and laying it' down upon Rodica's grave 'to protect her.' As we left we heard her mother, father, and grandparents singing, chanting, sobbing, heartbroken.

Later, we each branched out one by one to seek out Rodica's husband in the crowd and offer our condolences. I had brought semi-precious stones for her husband. I had carefully chosen them in a new-age shop, keen to offer something appropriate. I was very aware of having not collected them myself from a natural landscape, and it felt somehow not quite right. Nonetheless, each stone, apparently represented emotions or wishes for the widow: a stone to combat fear of the unknown, to aid in mourning, to offer strength … and I wrapped them in a brown silk cloth. This was a kind of self-made ritual symbol. An identity object that I had chosen due to what I knew of the couple. Rodica's funeral was the last time that I met with The Loose Community as a group.[17] With the help of ethnographic research carried out by Davies and Rumble[18] regarding natural burials, in Barton Glebe, Cambridge, I can maintain a certain compassionate and sensitive distance from direct critical analysis and instead approach this difficult day within the field as touching and insightful.

Identity Symbols and Performing Identity

> Many individuals possess deeply rooted symbols that are, practically, part and parcel of their identity and that families wish to incorporate within funeral or memorial rites [...] These things capture the meaningfulness of life, not least in its emotional tones.[19]

> issues of identity, control and choice are all deeply significant for natural burial for it is, at present, the quintessential expression of choice in relation to our bodies, albeit after death. [...] regarding our post mortal social presence [...][20]

The symbols within Rodica's funeral are suggestive of how she aligned herself with nature in life. Nature, here, is an 'identity symbol,'[21] one that was adopted as a social performance in remembrance of Rodica. The significance of the choices made for this day were due to how Rodica was to be remembered, but also how her identity may be aligned in death, in this moment with her relationship to the natural world. The symbols: the woodland, the wicker coffin, the flowers tossed, and the words spoken, were not only representative of Rodica's relationship with other humans but with the non-human world. These symbols present and actions performed by friends and family intended to verify who Rodica was in life: what she did, how she worked, her practice, her ethos, and her connection to nature. In death, Rodica's family wished to be faithful to her personality and did this through nature as a set of performative symbols. The event was framed, as any ritualistic performance rite would be. It was a temporal space to remember Rodica, to celebrate her life, but it was also the making of a new physical space where people may go to remember her—this space framed, separated, or made special by these symbols and acts.

To return to the earth in an animalistic manner, animalistic in that in the contemporary Western world, a natural burial is rare. This was a final performance of what Rodica considered in life—our interrelatedness with the organic. She would become a part, not only metaphorically but literally, of the organic network that makes up the forest. Rodica's life had been devoted to understanding our relationship to the natural, and a burial such as this allowed for further meaning making after death, providing a clear assonance between her lifestyle choices and beliefs and her 'death-style.'[22] The funeral was symbolic of her lifetime pursuits, her occupation, her relationship with her husband, her friends, The Loose Community, her goals, and ultimately her actual and ideal sense of identity:

> To locate one's authentic self in an authentic life is an ideal implicit in the desire to 'just blend back with nature'[23]

For Davies and Rumble,[24] these (arguably) romantic values are part of a contemporary trend 'towards the re-enchantment of the everyday world.'[25] Something, from within the book, we know this community to value. Rodica's burial then was very much in line with the community's tendency to animate the mundane and everyday. It signifies an appreciation of the organic processes, not as such a re-enchantment but instead an enchantment of the non-living as vitalised. Now not only the non-human but Rodica as well is a part of this network: this allowing Rodica a 'symbolic immortality.'[26] This allows the group and her family a sense of 'symbolic-continuity' (ibid). She will be here in and remembered through nature.

Binning Wood is now a place to walk for the community, acquiring its sense of place due to its acquiring Rodica,[27] enveloping her beneath its soil, transforming her into a part of this woodland. Her body, as before her mind, is now not only with nature but within—the ultimate self-verification in death and the ultimate solace, in loss, for those who perceive the world in a manner similar to Rodica. Binning Wood has become a part of the personal narrative of each member of The Loose Community. Through these performative acts, Rodica's husband, and Rodica in death have given this natural space a new context for themselves and the group.

Reflecting on the Book

This book aimed to answer several questions in relation to the activities that groups engage in when they journey to natural landscapes in Scotland and how people then evaluate the experience of these excursions and these spaces. It asked why some engage in these excursions, what it is that they do in these spaces, and what happens when they do. This considered emotional and behavioural aspects, group dynamics, and relationality. This book asked what within these excursions was considered positive? What do these excursions provide for individuals and groups? Why are these excursions to be transformative and how could they be considered as 'good' for individuals and groups? (Fig. 7.4)

The above diagram (Fig. 7.1) depicts the crux of my research findings, which I will reflect on now. I have shown that people join groups to make excursions into natural spaces due to a desire to belong and self-verify an aspect of their ideal or ought self in doing something that they deem to be *good* personally, socially, or environmentally. The definition of good has been something that has been left unsaid within this book. However,

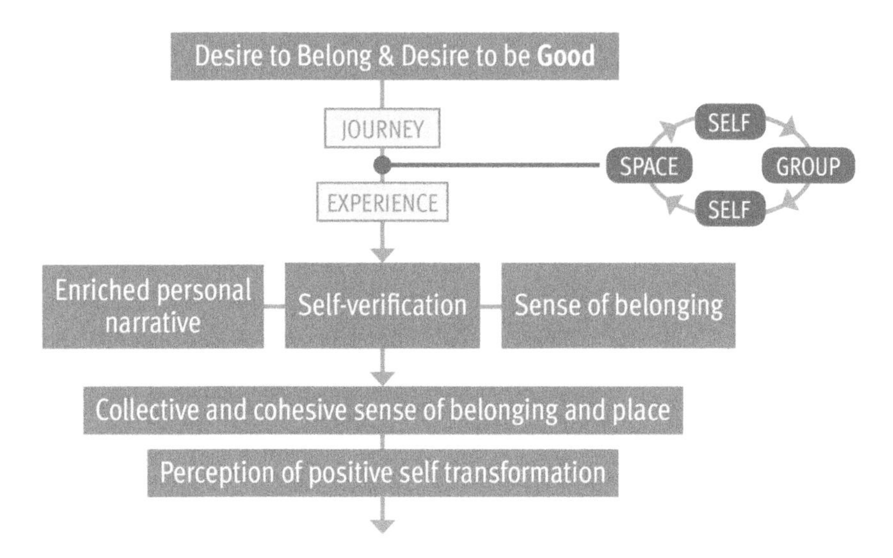

Fig. 7.4 Key influences on motivation and experience

we know that Individuals are seeking pleasure in these perhaps more abstract terms. A sense of belonging within a group and within a landscape as well as an enhanced sense of virtue brings pleasure in many subjective ways to my informants.

Within these journeys and temporal encounters, individuals and groups enter into what I call a *liminal loop*. These experiences are liminoid. The sense of self becomes a metaphorical liminal threshold, the group as a collective becomes a site of liminality as the dynamics shift and in turn affect the self and the group and so on. The group and individual perspectives of the landscape then become liminal—the threshold being somewhere between pre-context and post-context with regard to the way the landscape is perceived and how it relates to developing personal narratives. As this happens within the excursion experience, individuals build up new contexts, new associations of the landscape, and the social group within their own personal narrative. They verify an aspect of their ideal sense of self that they had sought by joining, they feel that they are doing *good*—perhaps leading to the belief that nature is *good* for us. Individuals feel a sense of belonging amongst the group and within the space. With this, a cohesive sense of group often emerges and thus there is a sense of positive transformation and wellbeing. I believe it is this that encourages individuals to repeat excursions such as this. These ideas are then reinforced with subsequent journeys.

Disciplinary approaches to nature connectedness however varied seemed to come to the similar conclusion that a connection to nature is something that is important, valuable, and positive to human wellbeing. Few influential theories explained the positive benefits of nature experience. Little research had sufficiently accounted for why being in nature was considered positive within the social sciences, particularly qualitatively. They did not account for the phenomenological experience of being within natural landscapes nor did they account for the relations between the self and non-human that are often ineffable. They did not explain fully why being in nature is experienced as positively transformative. This is what I aimed to do. The literature and empirical research was vast. As we have seen, the majority were quantitative. I, however have been concerned with the experiential, the phenomenological, the why, the personal, and the relational. I was interested in how these encounters are considered as transformative in relation to wellbeing. The research previous to my own did not seek to explain the qualitative experiential group encounter with natural landscapes. This is what I have done, for the case study groups with which I worked. My research has shown the qualitative ways in which some groups experience natural landscapes in Scotland. It has shown why some individuals are motivated to take part in activities within these spaces and how they are affected socially and emotionally by doing so.

To fully comprehend what happens to people when they journey as part of a group to natural spaces, it was necessary to gather experiential accounts. It was necessary for a researcher to understand the way that, relationally, the self, group, and landscape seemed to affect one another. It was necessary to consider how the group dynamic develops and how, not only people perceive the landscapes, but how they think about themselves and their own experiences within them. If I wanted to explore the self within these situations, I needed to explicitly look at identity. I needed to consider how individuals think of themselves within these encounters. It was within this that understandings of behaviour and experience began. One's identity encompasses social roles and geographical ties and so one's sense of self in relation to these spaces came into play. This became fluid within these spaces and as the group dynamic shifted. This became muddied when I considered that for some the social or interactive at least, was made up of not only the human. I have considered how the individual human, group and non-human, from the perspectives of informants, related to contribute to overall experience. I highlighted what lay between each that impacted these excursions.

In Chap. 1, I posed a question, perhaps controversial: do we appropriate people's aims for positive wellbeing to encourage pro-environmental care? And is this useful? By instrumentalising wellbeing, I wondered if, when the link is made between doing things in nature for wellbeing and the push towards doing our bit for the planet, this was helpful when addressing human wellbeing and mental health. For Adams,[28] pro-environmental care should be the motivation to connect with nature rather than encounters with nature being pushed for the sake of our own wellbeing. Instead it is pushed as a mutually reciprocal endeavour in which man might correct his destructive ways. I felt that this rhetoric does not aid in understanding human experiences within nature. For many, these excursions are about their own mental wellbeing and personal development, and it was this that I wished to explore. I have shown that regardless of agenda, individuals experienced positive transformations within these natural spaces by participating in group excursions. Rather than their wellbeing within these scenarios being instrumentalised for the sake of pro-environmental care, these groups instrumentalised the landscape in a bid to better their own personal circumstances. The motivations to 'connect' with nature were based in an endeavour for personal wellbeing.

In Chap. 2, I problematised notions of clearly defined methods within research of this kind—concluding that my methods would need to be, and were, more flexible, undefined, and spontaneous. It was here that I outlined my serendipitous and contextual ethnography within the field. I suggested that my methods may be closely aligned with Ingold's recent notions of 'anti-disciplinarity' (2018), and I outlined what it meant to *be there* in the field—socialising, integrating, journeying, learning and sharing skills, being silent, participating, being idle, and accepting contradiction. My methods were simple—go with, participate fully, listen, observe, and converse with people. By approaching the field in such a manner, liminality emerged as a prominent theme, where ambiguity, new spaces, and new people became of primary interest. Later, I came to the conclusion that these excursions may be closely linked to what Turner refers to as liminoid and van Gennep as liminal and that perhaps within this may be where we would discover communitas; perhaps the answer to the question of why these shared experiences were transformational.

Throughout Chap. 3, I detailed my time spent with groups and how methods became flexible and responsive to these groups and the research context. I spoke of the journey and the excursion and participation in activities as my primary means to understanding these experiences. I detailed the not instantly recognisable methods and why this serendipitous ethnography was a necessary framework for my means.

Throughout Chap. 4, I discussed my findings with regard to people's desires to find a sense of belonging within these excursions and an underlying want to verify a positive aspect of one's ideal or ought self. By taking part in activities within the natural that allowed informants' access to a new kind of cultural interaction, with new people and in new landscapes people began to form friendships with others who they deemed, for one reason or another, reflective of an aspect of themselves. This was in an empathic manner, within The Mental Health Initiative for example, and in the groups' ethos, or even the kinds of activities that they chose to be a part of. In these interactions with others and the space, people began to feel secure in finding a sense of belonging somewhere with others. Sometimes this was down to the social facilitation of the group. Given the many variables within these groups, the need to belong became a little more complex. It was not about belonging to any group. It was about becoming a part of a group that aligns with one's perceptions of oneself. For my informants, who all demonstrated a desire to be part of a group, their motivations were often articulated via more practical and less emotional reasoning, with the explicit need for belonging emerging as a by-motivation if you like. When I considered Baumeister and Leary's 'belongingness hypothesis,'[29] I was able to consider practical, emotional, and aspirational motivations as one and the same. By saying this, I suggested that regardless of specific motivation, a desire for belonging was at the centre. Ultimately, in this people verified an aspect of their ideal or ought self which brought about feelings of positive self and social transformation and a sense of wellbeing. This I have argued was due to three metaphorical liminal sites.

In Chap. 5, I laid out ideas regarding excursions as liminal—not only physical liminal thresholds but the sense of self as a metaphorical liminal site, the group dynamic as liminal, and the perception of the natural landscape as liminal. This I stated was to do with building on experience, personal narrative, associations, and then the perception of afforded opportunities for oneself and the group. Within this emerging of the liminal, I discussed the changing of structural norms, the desire for new structures, and for 'anti-structure' in some respects. This was aided in the field by the prominence of group reflection, for Turner, a specifically liminal realm. Throughout these excursions, as these thresholds were crossed whilst being a part of a group in the landscape, a sense of communitas, or group cohesiveness manifested: the groups went from being a group of individuals to a cohesive and bonded group—I became we.

Within Chap. 6, I broached the complexity of how people relate to the non-human landscape as either a mediator or intermediary in experience. Some viewed the landscape as having intention over their experience, and some believed in only the efficacy and sociality of the human within these encounters. This was about who or what had power over informants' experiences. With this came a complex spectrum, in which some anthropomorphised the landscape or believed in the sociality of nature, and some perceived only the human as having efficacy. I found though that this was not as straightforward as it may have first seemed, in that individuals, often unaware of their positioning could not be placed as either entirely ecocentric nor anthropocentric but instead somewhere in the middle—all groups used the natural landscape as a means to positive wellbeing.

In this conclusion, I spoke of discovering a certain vying for moral high ground across my case study groups. This was in terms of how one interacts with nature and concluded that all groups aimed to do *good* within the landscape regardless of definitions as to what *good* was. Ultimately, when people joined the groups to journey into natural landscapes and to verify an ideal or ought sense of self, it was because they deemed that activity or group to be good or to be doing good. I questioned what it meant to be *good* in nature. All groups believed that the way that they interacted with nature was the correct and *good* way to interact. The complexity of this was intriguing. My case study groups were aiming for some form of bettering, either in themselves, for others, or for the environment. Whether through aiming to understand the intrinsic value of the non-human, aiming to conserve and sustainably manage the landscape or aiming to better the lives of underprivileged youths, all relationships with nature were aiming to be *good*. The positive aspects of each of these pragmatic angles, often complex and undefinable, were difficult to dispute if we were talking of the aims of these groups. This was useful in understanding our complex relationship with the non-human and is something that may underlie all claims to positive self-transformation in the natural landscape—our moral positioning towards the non-human living natural landscape and what it means to be a *good* person. This brought us right back to ideas relating to self-verification and notions of the ideal self and what Cocks and Simpson referred to as 'satisfaction with oneself.'[30] One might argue that people wanted to be *good*. They wanted to associate themselves with acts and others who do good or who are considered favourably. It followed that the reasons for participating in excursions into the natural landscape was concerned with becoming or reaffirming oneself as a *good* person.

This brings us to where we are now. I have outlined what natural space represents for informants in terms of why they may choose to engage with these social and physical spaces. The motivation to engage with groups and in natural space appears to be due to a desire to belong within a given group and to verify some sense of the self in doing so. This ideal or ought sense of self was found by informants within the group's ethos and in the group's activities. This is often to do with subjective ideas of what it means to be a good person. We saw that the sense of self is enabled to metaphorically move closer to people's own perception of their ideal or ought self—a positive transformation. With this, the sense of self becomes a liminal threshold, and when each person is transformed along these lines, the group dynamics changes, and ways of perceiving (and thus behaving within and towards) the landscape is altered.

What '*Something*'?

My research began with the premise that people experienced an intangible 'something' positive when they journeyed with others into natural space. Whilst individuals sought 'something,' they also were aware that 'something' would and does happen within these encounters. This 'something,' was mentioned frequently throughout my fieldwork, and this 'something' was evidently different for each individual. However, this intangible 'something' has come some way to being exposed throughout my research. I believe that within these excursions it was to do with the relationality between the three subjects of self, group, and non-human landscape. This 'something' that people experience, I believe has to do with finding both a sense of belonging and sense of place, verifying an aspect of one's ideal sense of self and developing one's personal narrative in relation to place and group. I believe that this 'something' may be found in the crossing of those three distinct liminal thresholds. It may be found in exploring our own agency and perhaps the agency of others and the non-human. This 'something' may come down to feeling that one is a good person in terms of their relationship to the non-human natural landscape.

We need to go some way to discover more, with qualitative ethnographic methods, regarding the complex interactions between human nature and non-human nature. Perhaps this may follow the ethos of a serendipitous ethnography. These relationships are phenomenological and require methods that are capable of comprehending complex, contradictory data. Serendipitous, reflexive, and responsive methods may come closer to approaching experience which is often ineffable and intangible. Perhaps creative ethnographic methods may be developed to enable a more in-depth understanding of what happens within nature encounters. Creative methods may come some way in dealing with the implausibility of language often encountered in situations such as these. Creative ethnographic methods and an ethos of anti-disciplinarity might capture more of that ever elusive 'something' that occurs within human and nature relations.

With the above paragraph in mind, below are some suggestions for future research or simply questions to leave the reader with. These are questions that not only arose but have been left unanswered by this book:

What does it mean to be good in relation to human/ non-human relations? And, what does this have to do with our sense of who we are? How do we relate the non-human to ourselves?

Throughout fieldwork, individuals aimed to feel good and to verify a positive aspect of themselves. It would be valuable to understand how people consider *being good* or indeed being a good person in relation to human nature and non-human relations. It would be valuable to understand why and how we relate the non-human to our ideas of ourselves and why: is it possible to decipher whether our relationships and perceptions of the natural are innate or cultural? Could nature experiences be pleasurable with no social or cultural relevance? Could finding pleasure in nature be an innate primal reaction?

This is where I consider the dangers of overintellectualising activities that are potentially solely pleasurable for their own sake, though I am yet to be convinced this is possible. It is my belief that these excursions were pleasurable because of the effects on informants' sense of self (with social and cultural implications). It is worth considering whether in fact positive relations, indeed feeling pleasure, in nature could in fact be primal? Perhaps this could be based on gut feelings within places or groups of people. Perhaps by complying with environmental stimuli they may be built into personal narratives. Perhaps some driving motivations to be in nature are not in fact understood by the people themselves. Perhaps this could lead the way to answering another unanswered question: is there any plausibility for any other perspective than a human-centric one? Could our perspective be intuitive or instinctual? And finally, another rich area may be within the bounds of this question: how might we map the mind and body as a liminal threshold?

It would of course be possible to take notions of the sense of self as a metaphorical liminal site further. With more cognitive understanding perhaps, we may be able to understand the limits of both a literal and metaphorical liminal self. Might these questions be better understood through collaboration? To understand virtue qualitatively, to comprehend our primal relationships with nature, and to deal with the philosophical issues of where our mind resides, or how we may map mental thresholds and influences, we most likely will need to consider collaborative projects. These collaborative projects may utilise knowledges from the social and cognitive sciences as well as the arts and humanities. Perhaps this could be qualitative research carried out in collaboration with medical or neurological researchers. Or, alongside sociological, psychological, and anthropological researchers. Perhaps it may incorporate creative methods. We need to be open to transdisciplinarity: approaching from between and beyond disciplines. We need to remain open to new ways of knowing.

Notes

1. Cocks and Simpson (2015).
2. Cocks and Simpson (2015).
3. Hill Jnr (2006).
4. Hill Jnr (2006: 332).
5. Plumwood (2002).

6. Plumwood (2002: 181).
7. Hill Jnr (2006).
8. Hill Jnr (2006: 333–334).
9. Hill Jnr (2006).
10. Ferre in Attfield (2011).
11. John O'Neill in Attfield (2011).
12. in Attfield (2011).
13. Attfield (2011: 207).
14. Attfield (2011: 37).
15. This is this informant's real name. I made this decision alongside Helen and Rodica's husband as it felt more appropriate considering the circumstances.
16. Binningwood.co.uk
17. This final ceremony closed my relationship with this community (at least as a researcher.) Ethically, as an ethnographer and as a friend, due to the sensitive nature of this material, it did not feel appropriate to critically analyse Rodica's funeral nor the choices made by her husband. Instead this reflection should stand alone as an open comment regarding this circumstance and its relation to themes within my research. It is also with this realisation that I will refrain from obvious discussions of ritual and ceremonial, acknowledging that as events go, yes indeed death and performative rituals are so closely related to liminality. I will leave that here. I wholeheartedly wish to avoid accusations of ethnographic opportunism in the fate of stumbling upon death and funerary rites within the field, for the sake of my informants as friends and for Rodica's family. Besides, funerary rites of passage and traditional liminal thresholds are far from what my research thus far has been about. This is better left for elsewhere. Instead I will focus on the funeral as a series of symbols and performative acts representing Rodica's identity and her relationship with the natural world.
18. Davies and Rumble (2012).
19. Davies and Rumble (2012: 14).
20. Davies and Rumble (2012: 13).
21. Davies and Rumble (2012).
22. Davies and Rumble (2012).
23. (Davies and Rumble (2012: 68).
24. Davies and Rumble (2012).
25. Davies and Rumble (2012: 55).
26. Lifton in Davies and Rumble (2012).

27. A natural burial suggests a desire for independence from 'the 'unnatural interference of commercial ventures but an alliance with the organic world.' Whilst I could not comment on Rodica's beliefs regarding commercial venture, from what I know of Rodica and due to the circumstance in which we met, I feel most certainly that this kind of reasoning may have occurred to her. I am sure though, entailing many more ontological and spiritual complexities. Of course, losing the opportunity to speak with Rodica, I could never be sure of her motivations, or indeed her husband's motivations for organising the funeral in the way that they did. This may however suggest, for another researcher, further questions and areas of research that could be developed with regard to wellbeing and self-verification in death for both the deceased and family and friends of the deceased.
28. Adams (2009).
29. Baumeister and Leary (1995).
30. Cocks and Simpson (2015).

References

Adams, W. W. (2009). Basho's Therapy for Narcissus: Nature as Intimate Other and Transpersonal Self. *Journal of Humanistic Psychology, 50*(1), 38–64. https://doi.org/10.1177/0022167809338316.

Attfield, R. (2011). Beyond Anthropocentrism. *The Royal Institute of Philosophy Supplement, 69*, 29–46. https://doi.org/10.1017/S1358246111000191.

Baumeister, R. F., & Leary, M. R. (1995). The Need to Belong: Desire for Interpersonal Attachments as a Fundamental Human Motivation. *Psychological Bulletin, 117*(3), 497–529.

Binning Wood Memorial Forest. (2017). Available at http://www.binningwood.co.uk. Accessed 19 Apr 2017.

Cocks, S., & Simpson, S. (2015). Anthropocentric and Ecocentric: An Application of Environmental Philosophy to Outdoor Recreation and Environmental Education. *The Journal of Experimental Education, 38*(3), 216–227. https://doi.org/10.1177/1053825915571750.

Davies, D., & Rumble, H. (2012). Funeral Forms, Lifestyles and Death-Styles. In *Natural Burials: Traditional – Secular Spiritualties and Funeral Innovation* (pp. 1–18). London: Continuum. https://doi.org/10.5040/978147254910.ch-001.

Hill Jnr, T. (2006). Finding Value in Nature. *Environmental Values, 15*, 331–341. The White Horse Press.

Ingold, T. (2018). *Anthropology and/as Education*. Oxon: Routledge.

Plumwood, V. (2002). *Environmental Culture: The Ecological Crisis of Reason*. London/New York: Routledge.

Index[1]

[1] Note: Page numbers followed by 'n' refer to notes.

R. Crowther, *Wellbeing and Self-Transformation in Natural Landscapes*,
https://doi.org/10.1007/978-3-319-97673-0

R

S

CPI Antony Rowe
Eastbourne, UK
May 05, 2020